Happy CLIMATE—Happy LIFE

Be open to change to live in harmony with our planet

By
Michael Lohscheller
and **Ulrike Louis**

© 2024 Michael Lohscheller and Ulrike Louis

All Rights Reserved. Except for brief quotations included in critical reviews and certain other noncommercial uses allowed by copyright law, no part of this publication may be reproduced, distributed, or transmitted in any form or by any means, including photocopying, recording, or other electronic or mechanical methods, without the publisher's prior written permission. Write to the publisher at the address below, addressing your letter to the "Attention: Permissions Coordinator," requesting permission.

REVIEWS AND PRAISE

*"A complex problem clearly
structured and fun to read!."*
Dr. Tobias Tennstedt, CEO Trollkids

"This book invites humanity to take action to preserve Mother Earth by giving back more than we take out. Michael and Ulrike take the reader on an educative journey of well-laid-out facts that make the reader more knowledgeable. They pair this with sharing their personal journey of change, emphasizing that anybody can contribute in their own way. The seeds for change can be planted in many different ways, be it their teenage daughter, the automobile industry Michael has been leading as a senior executive, or simply daily life. The book aims at surpassing polarizing discussions and unites us in the belief that every step towards a more sustainable life will make our collective lives happier."
Joyshree Reinelt, CEO Innate Motion, Business Humanizer and Founding Partner

""Happy Climate, Happy Life" offers a compelling narrative intertwining practical sustainability insights with personal anecdotes, inspiring readers to embrace a harmonious relationship with the planet, and cultivating humanity's ability to collectively design a sustainable way of living for our future generations"
Brian Wennersten, Founder, Dutch Bamboo Foundation

TABLE OF CONTENTS

Foreword 1
Preface 4
Acknowledgements 7
Introduction 8

Chapter 1
First Things First 10

   1. The Facts 10
      1.1. Greenhouse Gas and Solar Energy Interaction 11
      1.2. Temperature Trends and Climate Change 11
      1.3. Human Impact on CO_2 Levels 12
      1.4. Climate Change Drivers 15
      1.5. Temperature Records and Trends 16
      1.6. Oceans and Sea-Level Rise 16
      1.7. Ocean Acidification and Sea Ice 16

   2. What's Being Done 17
      2.1. The Paris Agreement 17

   3. What Will Be Required From Us 18
      3.1. Planning 18
      3.2. Mitigation 19
      3.3. Adaptation 19
      3.4. Sacrifice 19

   4. Conclusion 20

Chapter 2
Living of the Future 21

   1. CO_2 of Living 21
   2. Housing 24
      2.1. Types of Housing 24
      2.2. Population Growth 25
      2.3. The Future of Housing 27
          2.3.1. Green-first Approach 27
          2.3.2. Materials of the Future 28

  2.3.3. Lifestyle 29
  2.3.4. Features to Add 30
  2.3.5. Sustainable and Eco-friendly Homes 31
3. Social 35
 3.1. Sharing 35
 3.2. Incentivizing a Smaller Family Size (Human Population) 39
 3.3. Work 40
4. Technology and Innovation 43
 4.1. Machines 44
 4.2. Energy 45
  4.2.1. Renewable Energy Sources 46
 4.3. Construction 51
 4.4. Sustainable Materials for Living 54
 4.5. Design of Cities 54
 4.6. Artificial Intelligence and Other Technological Advances 56
 4.7. Energy of the Future 58
 4.8. Carbon Neutral 59
5. Recommendation 59
 5.1. Evolve Our Value Systems and Economy 60
 5.2. Adopt Renewable (carbon-free/ carbon-low) Energy Solutions
  60
 5.3. Hire Green Builders 61
 5.4. Construct Eco-friendly Homes to Meet Global Demand 61
 5.5. Continue Professional Development (Building Experts) 61
 5.6. Construct New Features Using Eco-friendly Materials 62
 5.7. Use More Environmentally Friendly Appliances 62
 5.8. Go Green for Businesses 62
 5.9. Take Your Business to Green Enterprises 63
 5.10. Plant a Tree 63
6. Conclusion 63

Chapter 3
Food of the Future 65
1. What Is Food? 65
2. What Is Health? 67

3. C02 of Food 69
4. Diets 72
   4.1. Plant-based or Vegan Diet 72
   4.2. Vegetarian Diet 73
   4.3. Mediterranean Diet 73
   4.4. Ayurvedic Diet 74
   4.5. Ultra-processed Food Diet 74
5. The Humane Approach 75
   5.1. Cattle 76
   5.2. Chicken 78
   5.3. Fish 79
6. Food Waste and Food Security 82
7. How We Grow Food 83
   7.1. Local Food Supply Versus International Trade 84
   7.1. Regenerative Agriculture 86
      7.1.1. Agroforestry 87
      7.1.2. No-tilless Agriculture 88
   7.2. More Innovation 89
      7.2.1. Green Mobility 89
      7.2.2. CO2 as a Sustainable Input for Food Production 89
8. Holistic Wellbeing 91
   8.1. Linkage Between Diet/Lifestyle and Well-being 91
   8.2. Dopamine Management 91
   8.3. Taking Better Care of Ourselves 92
   8.4. The Placebo Effect 93
9. Recommendations 97
   9.1. Eat Predominately Locally Grown Food 97
   9.2. Support Local Farmers 97
   9.3. Reduce Food Waste: Eat "Funny" Fruit (and Vegetables) 98
   9.4. Eat Less Meat 98
   9.5. Lobby for Humane Practices in Farming and Less Food Waste 100
   9.6. Investing in Innovation 100
   9.7. Educate Others 101

9.8. Live a Healthy Lifestyle 101
10. Conclusion 101

Chapter 4
Mobility of the Future 103
1. What Is "Mobility"? 104
2. Classification of New Types of Vehicles 105
   2.1. Micromobility 105
   2.2. Powered Light Mobility 106
   2.3. Car and Van-like Vehicles 106
3. CO_2 of Transportation 107
4. Mobility in the City 107
   4.1. Ride-Sharing 110
   4.2. Fleet 111
   4.3. Collective Transport 112
      4.3.1. Popular Transport 112
      4.3.2. Public Transport 112
   4.4. Urban Development 112
   4.5. Expected Trends 113
      4.5.1. Decrease in Private Vehicle Usage 113
      4.5.2. RoboTaxis and RoboShuttles 114
      4.5.3. Increased Ride-sharing 115
   4.6. Regulations 115
5. Autonomous Vehicle (AV) 116
   5.1. Levels of Automation 116
   5.2. Where Are We Now? 118
   5.3. Advantages of AVs 118
   5.4. Disadvantages of AVs 119
6. Micromobility and Powered Light Mobility 121
   6.1. Micromobility Vehicles 121
   6.2. Micromobility Vehicle Types 122
   6.3. Advantages 124
   6.4. Disadvantages 126
7. Powered Light Vehicle (PLV) 129

    7.1. Advantages 130
    7.2. Disadvantages 130
  8. Intermodal Applications 131
  9. More Innovation 131
    9.1. Flying Cars 132
    9.2. Ammonia Engines 133
  10. Recommendations 133
    10.1. Changed Mindset 133
    10.2. Consider Total Cost of Ownership 134
    10.3. Shift to an Ecosystem Focus 134
    10.4. Carefully Plan Trips 134
    10.5. Use and Invest in Renewable-energy-powered Mobility 135
    10.6. Use Shared Transportation 136
    10.7. Intermodal applications may bridge the mobility gaps. 136
    10.8. Safety and Rules Will be Paramount for Mobility 136
    10.9. Use or Lobby for Green-Sources of Energy 137
    10.10. Lobby for Change in the Transportation Sector 137
  11. Conclusion 137

Chapter 5
Where Do We Need Help? **139**
  1. Economic Systems 140
  2. Incentivization 141
    2.1. Designing Incentives 142
    2.2. "Green" Carrots and "Green" Sticks 143
  3. Politics 147
    3.1. Carbon Tax 147
  4. Ecological 148
    4.1. Define Wildlife Areas 149
  5. The Experts 149
  6. Our Social Groups 151
  7. Recommendations 152
    7.1. Everything Needs to Have a Price 152
    7.2. Keep Things Simple 153
    7.3. Help From Experts 153

    7.4. Incentivize Key Stakeholders 154
    7.5. Teach Another 154
    7.6. Lobby our Leaders 155
  8. Conclusion 155

Chapter 6
Everybody Can 157
  1. What Are We Doing? 162
  2. Measuring Tangible Progress 163
  3. Learning From Each Other 165
  4. What Can We Start Doing Right Now? 167
  5. Recommendations 168
    5.1. List down what you are already doing to help stop climate change. 168
    5.2. Find out what your carbon footprint is 169
    5.3. Make a Carbon Footprint Goal 169
    5.4. Keep Learning and Teach Someone What You Have Learned 169
    5.5. Create Your 'Happy Climate List' 169
  6. Conclusion 170

Conclusion 171
Glossary 175
About the Authors 177
References 178

FOREWORD

During a lunch meeting in Rhede in early 2024, Michael told me that he and Ulrike reflected on one of the greatest challenges of our time - climate change and how to meet it. The couple realized that their careers, family life, and lifestyle provide unique lessons, worth sharing with their children and beyond. And so, they wrote it all down. This book comprehensively captures their research on climate change, wisdom from their personal efforts and aspirational intentions to reach a global audience.

Michael has contributed guest lectures to one of my courses at the University of St. Gallen in Switzerland. My students valued Michael's input and unique learning reflections as a top-manager. Within this book, it is in "only" one chapter, but many of the discussions with my students centered around the mobility of the future and it was fascinating how Michael shared his deep practical insights gained in many years of practical experience in the automotive industry. Topics such as how important it is to think of future mobility in conjunction with renewable energy sparked the minds of my learners.

So, when Michael asked me to write this forward, I immediately agreed. I think advocates for promoting sustainability in management and beyond are urgently needed.

Written in the Achterhoek, Netherlands - this book is a European contribution to an elite collection of literature for climate action that was written or edited by experienced practitioners and business leaders, such as John Doerr's "Speed & Scale", Bill Gates' "How to Avoid a Climate Disaster", Ayana Johnson and Katharine Wilkinson's "All We Can Save", and Paul Hawken's "Drawdown".

These written works complement the broad climate actions across firms and industries that an increasing number of leaders today set as strategy.— Books like this one are accessible to many, and thus provide unique potential to inspire debate on climate action.

Ulrike and Michael's personal style-of-writing in combination with their professional background provide powerful images and thus complement some of the existing more technical calls for climate action. In one story, Michael shares his thoughts on the sustainable shipping of running shoes which are important for him as an international marathon runner. Such stories shed a deeper light on his ambition and leadership as a former president and CEO of an entrepreneurial company aspiring to support sustainable road transport. The authors tell us how they power their electric vehicle with their own PV system at their private house or how they take their child to kindergarten by bike. It seems that Michael - with his quarter of a century top management experience in the automotive industry around the globe – found workable solutions to some of the most critical challenges of the mobility sector at the doorstep by trying and using them at home. In such, Ulrike and Michael offer their very personal solutions to addressing climate change.

The uniqueness of this book, "Happy Climate - Happy Life" lies not only in the personal stories the authors share but also in the positive attitude and advocacy for openness to change as a prerequisite for making any change. Reading Johnson & Wilkinson "All We Can Save" it becomes clear that how we think and speak about climate change and what we do matters. In times of increasing knowledge about the looming dynamics and consequences of climate change, people sometimes experience challenging emotions, such as fear or anger. The authors acknowledge the challenges but most importantly they establish a hopeful and constructive point of view that is worth emulating.

While comprehensive works can be overwhelming, there are often treasures of detailed insight to be found, as is the case of this book. Increasing climate action across industry and brave projects like this book can inspire fellow and future managers as well as people like me and you. I hope you find inspiration in "Happy Climate - Happy Life" and in the ideas that Ulrike Louis and Michael Lohscheller have collected and developed for you.

St. Gallen, Switzerland, in March 2024

Moritz Loock

PREFACE

Good or bad, change is inevitable. We all know this. Like you, we are no strangers to change. Whether we like it or not, life throws something at us that shifts the trajectory of our lives. How we respond to that change makes a world of difference. You know that this book is about climate change and we hope within its pages you will come to see why it's a topic so close to our hearts. For you to understand why we felt the need to write this book, we think it's important that we first share with you who we are and where we come from. Michael will begin.

I was born in Germany, close to the Dutch border, in 1968 to two hard-working parents. My father, a coal miner (eventually a teacher), began earning a living for his family when he was only just 15 years old. At the time, coal was a major economic driver for Germany. Though it was challenging work, it offered my father a steady income, which the family needed. As a result of the Second World War, everything changed for him. He suddenly found himself as the main breadwinner of the family. Later, he met my mother; they fell in love and they started their family. My mother supplemented the family income by washing curtains with my grandmother.

Of course, life got easier when my father became a teacher but we still lived in a degree of poverty. However, his shift to teaching helped instill in me a love for learning and an appreciation for teaching and sharing knowledge. As a result, I was fortunate enough to excel in school. Eventually, I landed tertiary education opportunities that changed the course of my life. Thereafter, I found employment that set me on a course of corporate success in leadership roles which I have been privileged enough to enjoy for over 25 years, primarily in the automotive industry. In addition, I was exceedingly lucky to meet and marry Ulrike and have our two children.

As my father taught me the value of sharing what one knows, this book is an attempt to do likewise. As you will learn through the pages of this book, Ulrike—my wife, paved the way forward for our family to make more environmentally friendly choices. Her influence, and that of my daughters, has helped us lower our carbon footprint by at least 40% over the past few years. It is this shared interest in addressing climate change that has led us to write this book in the hope of sharing what we've learned about climate change. For simplicity's sake, we decided to use the pronoun "I" throughout the book, except for this preface and each personal story (stories are in boxes) where we indicate whose personal experience it was specifically.

This book is for you if the concept of "climate change" is not a debate for you. It is not a thesis or scientific publication on the technicalities, theories, or contentions around global warming, climate change itself, and environmental degradation. Ideally, if you picked up this book, you are already convinced that climate change is a genuine problem that we need to address and fix on a global scale. Now, if you are not yet convinced that climate change is a real and significant problem, I hope that you are open-minded enough to entertain the possibility that there very well may be a problem that you are part of the solution for. If you have a desire to get practical and get involved in being part of the solution, then this book is definitely for you. Within its pages, we are confident that we will give you a good (if not comprehensive) starting point for your contribution to fighting climate change.

Science tells us that the temperature of the world's seas is rising and so are the sea levels. Weather patterns have become unpredictable in many parts of the globe leading to food insecurity, cyclones, and even draughts. We must stop what we are doing and start to remedy the impact that we have all had on the world.

This cannot be started tomorrow. We have to start now. I believe a good start to addressing climate change issues is getting ourselves educated about the topic, learning what can be done, and beginning to do what we can. We hope this book helps you do all three of these things. If we decide not to make any changes to arrest the damage done by our presence on the Earth, one day, we will have a different Earth, one that is hostile and inhospitable: an Earth with less land, less biodiversity, erratic weather patterns, and so much more that we can avoid entirely if we start to change. With that said, we hope you will allow us to share with you what we've learned about what is being done and what we can all do to change the world for the better. I hope you share my hopes for the future. Together we can drive the change that we want to see in the world. With that said, I invite you to flip the page and read on.

ACKNOWLEDGEMENTS

Writing a book is an exhausting endeavor. When we decided to take on this book, we had no idea what we were in for. The amount of research, debate and discussion, meetings, writing, and rewriting that this required would have been a daunting commitment to make, had we known what we would be in for. However, it's been an incredible journey that we have been privileged to complete.

Our gratitude goes to all the climate change experts and environmental enthusiasts who have generously shared their expertise and knowledge, which formed the basis of this book. The dedicated efforts of those addressing climate change were a constant source of hope and encouragement. Without them, we would not know the magnitude of the climate problems we face and the opportunity we have to change course.

Thanks goes to our two children, who are our biggest motivation for writing the book in the first place: We hope the example we live is embodied in this book. This is our gift to you and generations to come.

We would also like to thank our parents, friends, and family who not only encouraged us to write this book but were kind enough to read earlier drafts and give us feedback that helped us improve it as much as we felt possible.

Lastly, we would like to thank you: dear reader. This work would be a complete waste of time if you hadn't picked it up and thought it worthy of your time. Thank you.

INTRODUCTION

"... Every man has a right to his opinion, but no man has a right to be wrong in his facts...."
- Bernard M. Baruch[1]

We all enter the world through the same environment, no matter who we are. In that environment—our mother's womb—an egg is fertilized by our father. We experience, for the first time, what it is like to live in perfect harmony with our environments. Our mother's body supplies all the food and oxygen we need. There is even a waste management process to evacuate our waste. In this safe and miraculous place we sleep: we eat: we survive and grow for the next nine months. By taking care of herself, our mother take care of us. Then one day we exit the womb and enter the world.

Once we are out of our mother's womb, Mother Earth becomes something of a surrogate mother for each of us. As our biological mothers did, Mother Earth generously provides us with shelter, food, oxygen, and varying opportunities to live promising lives. But, there is one major difference. This difference is what compelled me to write this book. In our mother's womb, we lived a symbiotic existence, with a type of harmony. Now, with Mother Earth, mankind has lost grip of that harmony. There is no symbiosis. We have destabilized our environment to the point that there is no harmony between nature and man. One major sign of this lack of harmony is the enormous changes we see in our climate and the global environment.

If Mother Earth was a human being, I would say, our Earth is in a parasitic state. We have taken so much from her, and yet she still gives. A day may come when she may not be able to feed all of us or provide us with enough land to live on or even enough oxygen to breathe. As dire and maybe extreme as all this may sound, the science behind it is stark. We must heal Mother Earth and start giving back to her for all that she generously gives us.

What does giving back mean? That is what this book is about. I offer you ways that you and I, together, can start to create harmony with Mother Earth. What I offer you includes everyday lifestyle changes that you and I can make and what other stakeholders can do to take part in restorative efforts to address the climate change challenge. I believe that symbiosis between living things and Mother Earth is our birthright and something we can earn back. But, that is only if we are willing to make some sacrifices and changes together. This journey requires an open mind and acceptance that for the future to change, we must change.

Addressing the climate change challenge is not a job for one man, one country, or one interest group. No, it is a job for us all: one that our mother, Mother Earth, desperately needs us to take up. Are you willing to explore ways in which we can help her? Then read on.

Chapter 1
FIRST THINGS FIRST

"Climate change is the single greatest threat to a sustainable future but, at the same time, addressing the climate challenge presents a golden opportunity to promote prosperity, security and a brighter future for all."
Ban Ki-Moon, Former Secretary-General of the UN[2]

Some of the world's most credible scientists and renowned organizations have proven to us that climate change is real. They offer us facts that, if we are paying attention, can only lead us toward making sweeping changes in our lives. Through the following chapters, we will come across elements of our "carbon footprint" as individual contributors and as groups. A "carbon footprint" is the "total amount of greenhouse gases (including carbon dioxide and methane) that are generated by our actions"[3] Before we go into the chapters that address specific aspects of a more climate-friendly future, we first need to agree on two things: the facts and what these facts compel us to do. After exploring these two aspects, we will delve into each chapter and tackle one major aspect of human life on this planet after the other. So, let's agree on the foundation of it all: the facts.

1. The Facts

It is understandable if a few of us are skeptical about some of the information that you might stumble across over the Internet. Fortunately, many reputable organizations are offering scientifically backed resources explaining the length and breadth of the challenges at hand when it comes to climate change. In this section of this chapter, we want to spend a moment looking at some of the most devastating and reputable aspects of climate change that we cannot afford to ignore. To make our exploration easier to digest, I've grouped this information into categories that I think help us make more sense of it all. These categories are:

- Greenhouse gas and solar energy interaction
- Temperature trends and climate change
- Human impact on CO2 levels
- Climate change drivers
- Temperature records and changes
- Ocean sea-level rise
- Ocean acidification and sea ice

Let us begin to look at each.

1.1. Greenhouse Gas and Solar Energy Interaction

GHGs or greenhouse gases are gasses in the Earth's atmosphere that act like the glass of a greenhouse and keep the Earth heated. As a result of humanity's impact on the Earth, the increased release of greenhouse gasses has been massive, leading to what we know as global warming and climate change[4]. According to NASA, approximately 70% of the sun's energy that hits the Earth is absorbed and released but some remains trapped by gasses[5]. Some of these gasses causing this warming of the Earth, known as the greenhouse effect, include carbon dioxide (CO2), methane, nitrous oxide, and water vapor[6].

1.2. Temperature Trends and Climate Change

You will see us discuss carbon dioxide extensively in this book because of its relevance as a greenhouse gas devastating the global climate. Data shows us that CO2 has increased by over 50% in under 300 years[7]. In the timing of climate change, 300 years is a short time which makes the increase quite alarming.

What is additionally interesting is that the sun itself seems to be cooling down, whilst the Earth heats[8]. At the same time, the bottom part of our atmosphere is warming while the atmosphere above it is cooling[9].

Warming of the Earth is not a new phenomenon. The Ice Age ended because the Earth started to heat up. The uncomfortable fact though is how much faster the Earth is warming up. This is at a speed that's almost 10 times faster than that of the Ice Age[10].

1.3. Human Impact on CO2 Levels

A range of human activities cause the release of additional CO2 into the atmosphere. These activities include the burning of fossil fuels, deforestation, changing the use of land, industrial processes, waste management, agriculture, and some natural processes like erupting volcanoes.

In a 2022 report by the International Energy Agency (IEA), energy-related CO2 emissions increased by 321 Mt, representing a 0.9% increase[11]. Furthermore, the overall CO2 emissions reached a new global high of 36.8 billion tonnes. According to the report, the biggest contributor to CO2 emissions was from coal-fired electricity and heat. Regardless of this new record, the projection anticipated as a result of worldwide economic growth was worse.

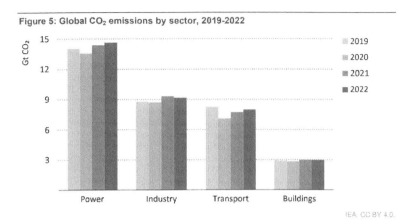

Credit Source: International Energy Agency [12]

The sectors that this diagram presents are power, industry, transport and buildings. "Power" includes the combustion of fossil fuels like coal, natural gas, and oil. "Industry" refers to the global contribution of gas emissions from the combustion of fossil fuels on commercial premises for energy generation. This encompasses metallurgical, chemical, and mineral transformation processes not directly resulting from energy consumption and emissions from byproducts. "Transportation" refers to emissions produced through burning natural resources for fuel for all types of vehicles including trucks, cars, trains, airplanes, and boats. "Buildings" refers to the contribution of GHGs onsite energy generation and combustion of fuel for heat in buildings or cooking in residential establishments.

As you can see from the chart, a major source of CO_2 and other greenhouse gasses is the burning of oil, natural gas, coal, and other fossil fuels for energy required for our mobility, power, and industrial processes.

Although not depicted in the diagram, other factors like land use, deforestation, construction, agricultural practices, and food production also contribute to CO_2 emissions. Deforestation contributes 12–20% of global emissions of GHGs [13].

Plants and trees absorb carbon dioxide, nutrients, and water to synthesize food using sunlight through the process called photosynthesis [14]. This process is good for us in lowering the amount of GHGs in our atmosphere and avoiding the increasing problem of global warming. Plants also "sweat" to cool themselves, and by extension, their surrounding environment[15]. However, dead and burnt trees release that carbon dioxide into the atmosphere.

Then there is the matter of CO2 from the food production process. CO2 is expanded at various levels of food production, from land use to the animals being farmed for food and the transportation required to get the food to the consumer.

Researchers estimate that approximately 20% of global GHGs are from simply getting the foods to us, the consumers[16]. Findings from 2018 attributed 17% of global GHGs to agriculture and land use[17]. Getting that piece of steak, leg of lamb, brisket or meatloaf involves raising livestock that happen to have digestive systems that release significant quantities of CO2 as a byproduct of consuming feed. Other elements of crop production, such as using chemically manufactured fertilizers in turn contribute even more CO2.

The figure below from the Food and Agriculture Organization (FAO) of the United Nations depicts data on global emissions from agriculture and related land use and presents it in relation to overall global emissions from the years 2000 to 2018.

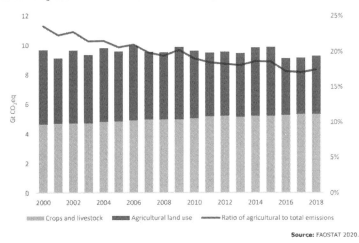

Figure 1. Yearly emissions from crops and livestock and related land use, and share of agriculture in global GHG emissions from all sectors, 2000–2018

Credit Source: FAO. 2021[18]

We will take a more comprehensive look at the impact of our food in Chapter 3.

When land once left fallowing or rotating in usage for planting crops or raising livestock, is converted to a property for commercial or residential usage, carbon stored in soils and vegetation is let into the atmosphere. Other than this, industrial processes like cement production involve the release of even more carbon into the atmosphere. Landfills are not just eyesores. They also release CO2 emissions as organic matter decays.

1.4. Climate Change Drivers

Before humanity was such a formidable force in contributing to climate change, volcanoes were among the biggest drivers of the increase of greenhouse gasses around the world. They not only released CO2 into the atmosphere but water vapor, sulfur dioxide, hydrogen halides, and hydrogen sulfide. All of these gasses, except for water vapor, can be detrimental to the health of living things that are exposed to them[19] . Though volcanic eruptions are hardly a common occurrence these days, the volume of such gasses released in an eruption is significant. For context, consider Mount Pinatubo which released over 250 megatons of gas into the upper atmosphere in just 24 hours[20].

1.5. Temperature Records and Trends

Did you know that the year 2023 was among the warmest in recorded history? It should not surprise us then as studies also show us that the Earth's temperature has gone up every 10 years at an average rate of 0.08 degrees Celsius or 0.14 degrees Fahrenheit[21]

You may also find it alarming that of all the 10 hottest years on record, all happened after the year 2010 [22]. This alone should cause us to pause and reflect. What shall we expect next? As trends go, we cannot reasonably expect this trend to change shortly.

1.6. Oceans and Sea-Level Rise

If you have to guess, where do you suspect the majority of the Earth's heat ends up? Well, oceans. These epic bodies of water absorb a whopping 90% of Earth's heat! [23] The Earth's atmosphere only retains a meagre fraction (2%) of the Earth's heat [24].

Sea levels have risen ~8 inches since 1901, and are rising 0.1 inches per year. Research tells us that now not only are sea levels rising, but at an even faster rate[25]. The Earth experienced little change in its sea levels between 0 and 1900[26]

1.7. Ocean Acidification and Sea Ice

Acidity is the concentration of hydrogen ions in a solution. CO2 dissolving in the ocean drives ocean acidification. As it stands, not only is the acidity of seawater increasing because of us human beings but this rise will impact and change the ecosystems in our oceans and seas, negatively impacting the biology of the diverse living creatures. It took millions of years for the Earth to regain much of its biodiversity[27].

2. What's Being Done

Governments and organizations around the world recognize the severity of the problems that climate change poses for all of us. With that understanding, many initiatives and efforts are being made to attempt to address it. One of the most important global initiatives worth mentioning here is the Paris Agreement.

2.1. The Paris Agreement

This is an international agreement entered into by 196 countries in 2021 in Paris, France. The intent of that agreement was for nations to collectively work together [28]. Some of the key elements of the agreement included a long-term temperature goal, global peaking and "climate neutrality," mitigation, sinks and reservoirs, voluntary cooperation, adaptation, loss and damage, finance, technology and capacity building support, climate change education, transparency, and global stocktake[29].

The highest goal of this agreement is to curb the rise of the Earth's temperature and keep it at temperatures below 2 degrees Celsius—the rise the Earth experienced before the industrial levels. However, more recently global leaders have further adjusted their target to below 1.5 degrees Celsius[30]. This is because of the potential for a rise above 1.5 degrees Celsius to lead to more frequent and more devastating unpredictable climate disasters like droughts, torrential rainfall, and heatwaves.

Sadly, regardless of these public and international commitments, the results so far are less than encouraging. Of course, this is challenging to measure[31]. The United Nations warned global leaders that the world was still not on track to meet the Paris Agreement and would need to aim higher if we are to experience the results the

Earth desperately needs[32]. From a diplomacy perspective, countries may agree to treaties of this nature but fail to genuinely produce results. In other cases, some countries even make pledges that are considered significantly unimpactful. Countries like Saudi Arabia, Russia, and Iran are among those criticized for demonstrating highly insufficient commitment[33]. Ultimately, the Paris Agreement and agreements like it cannot truly be enforced, and that adds to the core of why these alone cannot solve our climate change problem.

3. What Will Be Required From Us

The facts presented above have painted a bleak future for our Earthly home. Thankfully, there are measures you and I can take. This is not a problem to be left to our leaders and organizations to rectify. This is a problem that intimately affects each of us; as such each of us needs to be part of the solution. For that reason, the rest of the chapters of this book will attempt to look at the efforts you and I can make. For now, though, it would be a great time to understand what the rest of this book focuses on: what we can do. Briefly, we'll take a look at the following that can help us manage the issue of climate change: planning, mitigation, adaptation, and sacrifice.

3.1. Planning

As individuals, and as a collective of the human race, planning can help us ensure we win against climate change. Being afraid or saddened by the impact of climate change is a necessary part of changing our behaviors but it isn't enough. We need to plan our way out of this dire situation.

Every planning cycle needs clear targets. Every target needs to be followed up and measured. Deviations need to be shown and actions need to be defined to mitigate them at the global, national, and local levels. As it stands, our ability to beat climate change falls short at all levels, beginning with the very basics: effective targets, measurement of results, and sustained execution of climate-friendly efforts. Without a serious evaluation of all these components, at best, the global efforts put forward are nothing more than a diplomatic exercise with no real substance to it.

3.2. Mitigation

This is a strategy that involves avoiding and lowering the amount of greenhouse gasses that we release into the atmosphere[34]. Mitigation may take years or even decades for any of our efforts to finally bear fruits.

3.3. Adaptation

On the other hand, "adaptation" can be defined as follows:

> "Climate change adaptation means altering our behavior, systems, and—in some cases—ways of life to protect our families, our economies, and the environment in which we live from the impacts of climate change."[35]

3.4. Sacrifice

Nothing worthwhile ever just happens. It requires work, effort, and a genuine desire to see the world going in a more life-affirming direction. If it's that important to you and me we'll have to marry our planning, mitigation, and adaptation efforts to one more important aspect: sacrifice.

Sacrifice may look different for everyone but our collective efforts are what shift us into a new reality where humanity thrives and so does its environment.

4. Conclusion

There are many other facts about climate change that we have not looked into further. However, the facts we looked at should be enough to convince you that we don't have any time left to make half-hearted efforts to address climate change. Now that we're done with these facts, let's jump right into the solutions. Let's begin by looking at how we will get around in a climate-friendly future. In the next chapter, we will take a more comprehensive look at how we live and what can be done to mitigate or adapt to climate change in our daily lives.

Chapter 2
LIVING OF THE FUTURE

"Yes, but I have something he will never have... enough."
Joseph Heller[36]

Existing on planet Earth has its costs for each of us. No matter what we do, we make an impact on others and the environment. In this chapter, we delve into what the future will need to look like for us to truly have brighter futures. We will begin by looking at the CO2 of living and then address the living-related areas of our climate impact such as housing, social impact (our clothing, work, human population, and other civilization-related elements that cost our environment), technology, and innovation (energy, construction, and design of our cities).

1. CO2 of Living

On average, each of us is responsible for emitting 4.7 tonnes of CO2 and other GHGs into the atmosphere per year[37]. The variances between usage on a socio-economic and geographic level are significant, with average yearly CO2 emissions per individual in the West and other highly developed regions (such as the Middle East) being staggeringly higher. For example, in the United States, an individual emits 16 tonnes of CO2 per year[38]. In 2022, Qatar's per-capita CO2 emission was 9.4 times more than the global average at 37.4 tonnes per person. Here's a map that shows us the average carbon footprint per person based on country:

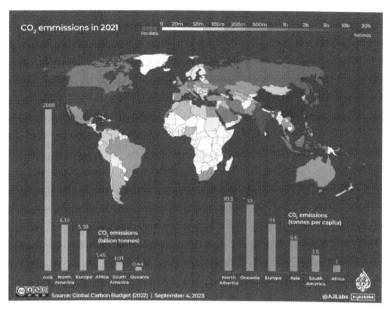

Source (cropped): Al Jazeera, 2021 [39]

What may surprise you is that the most affluent—global 1%—produced over 1000 times the CO2 emissions of the world's lowest economic earners[40]. If we look at this further, we see that the most vulnerable countries that will experience climate-related disasters are in majority, those that do not contribute as much to global CO2 emissions. According to aid organizations (International Rescue Committee and World Resource Institute), the countries most at risk of climate-related disasters include Somalia, Syria, The Democratic Republic of Congo, Afghanistan, Yemen, Chad, South Sudan, the Central African Republic, Nigeria, and Ethiopia[41]. Climate-related disasters that some of these countries have and may experience include earthquakes (as with Syria earlier this year), torrential rain (as with the DRC), ongoing drought—worst in history (Afghanistan), food insecurity (Yemen), and flooding (Chad, Central African Republic, and Nigeria)[42]

In 2021, statistics showed that the buildings and construction sector was responsible for 34% of global energy demands and close to

37% of energy- and process-related CO2 emissions as shown in the graphic below:[43]

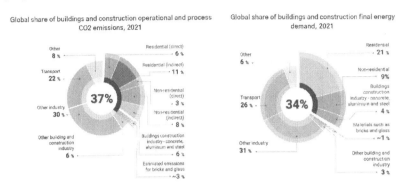

Source: UN Environment Programme - —Global Alliance for Buildings and Construction[44]

In the graphic above to the left, you may note the global share of CO2 emissions related to buildings and construction. To the right, we see the energy demand for the same period: 2021.

We would all like the world to be free from poverty. The unfortunate matter is that global GDP is linked to 99% of energy consumption and 100% of material uses[45]. This is because thriving economies have higher energy needs to meet. This means, the wealthier we become, the higher our energy needs. Forecasts predict that our global energy and material needs will double in the next 25 years[46] This is all thanks to our collective need for comfort, especially in our built environments (man-made surroundings).

Desire, a European Commission's New European Bauhaus lighthouse project stated the following in their 2023 paper inviting dialogue related to the living of the future, particularly about our built environment and our predisposition for comfort:

> *"Reimagining comfort in the 21st century new Bauhaus economy requires us to firstly recognise we are complicit in the vast injustice of comfort, driven by a stream of micro violences in our everyday lives: noise pollution, air pollution, light*

pollution, detachment from our natural ecosystems and habitats. Further, we have to acknowledge how we achieve comfort for some in the built environment through the significant allocation of services and technologies places great demands on energy and materials, generating significant waste and impacting current and future, human and non human generations. We will need to both change how just comfort is achieved and delivered and reexamine the actual provision of comfort for the 21st century." Dark Matter Labs. 2023[47]

2. Housing

All of us live in some sort of housing. Due to natural disasters, wars, and other extreme circumstances, some of us may be forced to use temporary shelters for accommodation. Homes emit some form of emissions as they are constructed and as we dwell in them through heating and electricity[48]. The United Nations Environmental Programme states (as of 2022) that the buildings and construction sector (which heavily influences the CO_2 of living) is not on track to achieve decarbonization by the year 2050[49]. I would like us to look at several factors of housing and buildings in general, including; types of permanent accommodation, materials utilized to build them, and the direction we need to take going forward, for the benefit of our climate. We will begin by looking at types of housing.

2.1. Types of Housing

Apartments, cooperative housing, condominiums, manufactured houses, and single-family houses are all types of houses that are available to many of us around the world[50].

A "single-family" home is permanent accommodation meant or typically designed for one family[51]. They are freestanding and often standalone. An "apartment" is, *"usually (a) rented room or set of rooms that is part of a building and is used as a place to live"*[52] An individual apartment unit is often on one floor, typically found in cities, and often utilized on a rent basis[53]. A "condominium" or "condo" is a type of residence typically in a large complex containing numerous individual dwelling units, with each unit being individually owned[54]. Known as a more affordable type of housing than conventional single-family houses, "manufactured houses"—previously popularly known as "mobile homes" are built in factories, piece-by-piece, and are assembled where the home-buyer wants to call home[55]. As with a traditional home, some homeowners may feel the need to take out a mortgage to make this homeownership a reality. "Cooperative housing" is a type of residential housing where the owners are shareholders and residents of the units that they collectively own[56].

Regardless of the type of housing we live in, the building has had an impact on the environment throughout its existence, otherwise known as a "life cycle." A building's life cycle has the following stages: the construction stage, usage stage, end-of-life stage, and externalized impacts beyond the system boundary[57].

2.2. Population Growth

Our global population is a big factor in the amount of CO2 that our living contributes. The global human population reached 8 billion in 2022, and the United Nations projects that it will reach 10 billion over the next 30 years[58].

As our population has increased, our household sizes and related trends have also changed globally. Over the last 50 years, the average household sizes (though there's a significant variance worldwide) show a decline[59]:

> *"For example, in France, the average household size fell from 3.1 persons in 1968 to 2.3 in 2011, the same time the country's fertility rate fell from 2.6 to 2.0 live births per woman. In Kenya, the average household size fell from 5.3 persons per household in 1969 to 4.0 in 2014, in line with a fertility decline from 8.1 to 4.4 live births per woman." The Conversation, 2018[60]*

This population growth, ageing populations especially amongst developed nations, and smaller household size have also brought with it a growth in the number of one- and two-person households[61]. The infertility rate is on the rise in various parts of the world such as North America, Europe, China, and Brazil[62]. This correlates with these smaller household sizes. Africa and Asia generally retain their tradition of having multi-generational homes.

Statistics also show that household demand is rising faster than actual population growth[63]. Commentators state that if this rising housing demand continues globally at a rate of 7–8% for the next 80 years, we will need approximately 800 million new homes to house everyone adequately[64]. The Conversation also adds: *"Taking an average global three-person household (1.2 billion homes) coupled with that 8% demographic factor of total global population over the period results in a need for more than two billion new homes by the end of the 21st century."*[65] Demographers have dubbed the event of the realization of such an epic human population the "population bomb" and the environmental impact that this will have on our climate would be of unprecedented proportions:

> *"That means the same number of people today live in twice as many homes, requiring twice as many resources to build and furnish them, to heat and cool them, and to pave roads to their*

front doors. This "household explosion" has long been underway in developed countries. But it's rapidly accelerating throughout the rest of the world." Bloomberg. (2014)[66]

Your location and the size of your home are both factors in the amount of carbon your home emits. One particular study conducted on American homes showed that high-income homes utilized more energy than poor homes[67]. That energy, without widespread usage of clean energy, translates into more GHGs in our atmosphere.

One need that contributes significantly to CO2 emissions in the home is our need to heat our homes. Particularly where the temperature drops below inhabitable levels, furnaces or boilers stocked with fossil fuels are typical[68].

2.3. The Future of Housing

Housing of the future that's affordable, climate-friendly, and sustainable is the ideal. "Sustainability", in this context is "the ability to maintain or support a process continuously over time. In business and policy contexts, sustainability seeks to prevent the depletion of natural or physical resources, so that they will remain available for the long term."[69]

2.3.1. Green-first Approach

Homes and other structures for the future will require a green-first approach at all levels: planning, design, construction, and usage or operations[70]. Some services and apps can help us achieve sustainable housing, such as SHERPA, an app by the UN-Habitat[71]. SHERPA and other apps like it offer a well-rounded approach to assessing the sustainability of a housing project; SHERPA in particular includes cultural and economic factors, as well as aspects of social sustainability[72].

2.3.2. Materials of the Future

Have you ever heard of composite materials before? These are materials made from more than one material which subsequently has greater properties as the combined product[73]. Eco-friendly composite materials can be used instead of more conventional materials without sacrificing quality and may offer several benefits such as lightness, resistance to corrosion, high mechanical resistance, malleability, tailor-made finishing, and ease of maintenance or reinforcement[74]. The increased usage of green composites will help us fix the climate problem and if plastic waste is also utilized, it will help us recycle non-biodegradable plastics through their incorporation[75]. Examples of eco-friendly composite materials include cork fiber and cardboard waste[76].

Some examples of sustainable construction materials include bamboo, hempcrete, recycled steel, rammed earth, cork, recycled glass, straw bale, earth blocks and reclaimed wood[77]. Bamboo is technically a grass that looks like a tall slender tree. Many varieties of bamboo typically grow fast[78]. Hempcrete is a "material made by wet-mixing hemp shiv (desiccated woody core of the plant) with a lime binder. It provides a vapor-permeable, airtight insulating layer which also has great thermal mass"[79]. As steel is a non-renewable construction material, recycled steel is one way to still utilize steel that has already been utilized in a construction project. Recycled steel does not lose its properties and is therefore just as strong as it was the first time it was made, and happens to be the most recycled material on Earth[80]. Rammed earth is a material made by compressing select soils for building and can be utilized in the form of blocks or even whole walls put in place as layers[81].

Cork is matter harvested from the outer bark of the cork oak tree. Just as with steel, glass can be endlessly recycled without any degradation of quality. Using recycled glass in place of creating brand-new glass is climate-friendly for the following reasons:

> *"Over half of the energy consumption of the glass industry is used for melting to form the glass. Adding recycled glass to the raw materials reduces energy use and CO2 emissions. Another advantage is that less raw material is needed. Currently, the world average glass recycling rate is about 50%. Higher recycling rates are possible, especially in regions where the recovery rate is still low..." UN Climate Technology Centre & Network. (n.d.)[82]*

Similar to how rammed earth is produced for construction purposes, straw bales are compacted to create a building material that can be either load-bearing (structure of the wall) or non-load-bearing (filling cavities in the wall)[83]. Straw-bale construction, as with rammed earth construction, has been done for thousands of years in places like the Middle East and shines as an eco-friendly construction option today[84].

Reclaimed wood is wood that has been used previously for various purposes such as building or structures (e.g., barns, industrial buildings, water tanks, gym bleachers, ships)[85]. This material is more environmentally friendly than recycled steel as the recycling process for steel requires the use of fuel (typically fossil fuels), therefore emitting CO2 and other GHGs into the atmosphere[86]. Though chopping down our forests to produce wood for our homes is not environmentally friendly, using reclaimed (recycled/salvaged) wood is. Reclaimed wood beats steel and concrete as the more eco-friendly option of these most common building materials.

2.3.3. Lifestyle

Apart from a green-first approach for the construction of our homes and even other types of buildings, we also have an enormous opportunity to support a healthier climate by making lifestyle changes. Adding climate-friendly features to our existing homes, opting for sustainable & climate-friendly new homes, and sharing accommodations are some of the lifestyle changes we can make. Let's briefly look at each of these options.

2.3.4. Features to Add

Features that allow you to use renewable energy such as solar power are an excellent addition to your home, as they are not only climate-friendly but can be cheap[87]. More comprehensive systems can have a pricey upfront cost but will save you a great deal of money in its lifetime of usage. Energy-saver systems exist that allow you to set the wattage you require at a specific time by using timers, dimmers, and various smart lighting solutions[88].

We have already talked about sustainable and eco-friendly materials for the construction of new buildings. We can still reap some benefits for the environment by incorporating eco-friendly materials for extensions or new features to our existing homes with the help of building professionals who are skilled at working with these more climate-friendly materials like salvaged wood, recycled steel, recycled glass, hempcrete or straw bales.

Home gardens are another eco-friendly feature worth setting up if you don't already have one. Methods like permaculture not only add to biodiversity but support the local ecosystem, without taking away from it[89]. Permaculture is an innovative agricultural practice for sustainable living that practically encompasses developing harmonious, efficient, and productive systems for a range of environmental contexts[90].

Proper insulation, smart thermostats, and Energy Star appliances are additional features your home and the environment can benefit from[91]. Doors and windows that don't close properly cost you more for heating, particularly in colder times of the year. If you live in the global North, or other parts of the world that experience extreme cold, this may add significantly to your annual utility bills.

2.3.5. Sustainable and Eco-friendly Homes

Prefabricated homes, tiny homes, earthships, LEED-certified homes, solar-paneled houses, biophilic homes, earth-covered homes, and passive homes are all options for sustainable and eco-friendly homes[92]. Previously referred to as mobile homes, prefabricated homes are factory-made houses that are made in parts and assembled at the site[93]. Tiny homes, or small houses, have a floor plan of 500 square feet and are sometimes prefabricated[94]. Lowered labor costs for the home buyer and less material waste are among the benefits of prefabricated homes[95]. Have you ever heard of earthships? An *earthship is another type of sustainable home. They are "living buildings that are adapted to their environment to make use of climate, water, and waste products. When people think of earthships, they usually think of the desert southwest of the US, because that's where they originated and where many of them are built."*[96]

In the US, you have the option of purchasing a LEED-certified home. This is a certification awarded for buildings that reach or exceed the standards set. This is a United States Green Building Council award for "leadership in energy and environmental design[97]. The perks of buying homes which are certified include lower maintenance, and a comparable cost to non-certified homes but offering savings to the homeowner of up to 30% on utility bills, making these homes the savvier choice[98].

When we talk of a solar-paneled house, you can have a hybrid system (using both electricity from the grid and your solar-generated electricity) or a solar-only system. If you already have a home, attaching a solar system may be the more cost-effective option of the two. However, if you are constructing one, installing building-integrated photovoltaics (BIPV) may be the more seamless option to directly provide electricity to your home.

Solar Panels are "Ugly"!

"I'd put my money on the sun and solar energy. What a source of power! I hope we don't have to wait until oil and coal run out before we tackle that."
Thomas Edison[99]

When my family and I moved house, we were excited about finding our new home. The house-hunting was exhausting but we finally laid our eyes on the house we knew was a great fit for our small family. The house we found, which became our home, was nestled in the beautiful countryside. The lush green landscape in spring, the clean air year round and the birds chirping every morning delight us. But do you know what one of the major selling points was for the house we ended up selecting? The inbuilt solar system! The house already had one installed. At the time, my family and I were already passionate about caring for our planet, and now we are even more in love with our home.

In most seasons, our solar panel home does a great job. With it, we can power all the appliances we use. We recently bought an electric car, to use alongside (and eventually in replacement of) our beloved car that has an internal combustion engine. Winter (December through February) is a different story though. Since days are short and nights may last up to 17 hours in winter over here, there are simply not enough hours in the day for our system to generate the *power we need[100]. That is even without including our electric car's energy needs in those months. As a result, we use power on the grid to help us meet any of our unmet energy needs. We would prefer if this was not the case.*

Many households where we live produce considerable energy surplus which they send back to the grid and receive credit for. The population is so good at this that the government is considering lowering the credits we receive because it's costing the government 400 million euros in lost revenue taxes[101].

Our family's carbon footprint is far lower by using solar power for most of our daily energy needs around the home and even our cost of mobility—because of the electric car. We know that going fully off-grid makes the most sense for us from many different perspectives.

Ulrike

Biophilic design is another type of eco-friendly home. This is a home design characterized by a style that "connects homeowners to nature with elements such as indoor plants and fountains, terraces, and gardens. Views of the ocean, mountains, and other outdoor landscapes also figure."[102]

Because these home designs are crafted with nature as an important component, residents of such houses reap the benefits that nature offers such as a positive impact on their mental and physical health, and their overall wellbeing[103].

Have you ever watched any of the Hobbit movies or read the books that these movies were based on by J.R.R. Tolkien?[104]? In it, the main protagonists dwell in "hobbit houses," a type of fictional home that also happens to be a kind of earth-covered home. Though the movie's versions are fictional, real-life earth-covered homes that these are based on can be found in New Zealand and Iceland[105]. If you enjoy TV shows, you may have come across some of these earth-covered homes in Game of Thrones, where some scenes were shot in Iceland[106]. Earth-covered homes or earth-sheltered homes are houses that are constructed with a full or partial layering of earth, on a flat site, with the main living areas surrounding an outdoor courtyard[107]. The main advantages of these types of homes are their temperature control characteristics: extreme weather will not cause major issues inside the house[108]. With climate change causing natural disasters around the world, the destruction of property, including homes is not uncommon. This is where this type of home comes out on top! These homes give you additional protection against some natural disasters and high winds[109]. From a financial perspective, they are a more economical option as insurance costs less than that of a conventionally designed home.

A passive house is the last sustainable and eco-friendly home design that we will look at here, even though other options certainly exist and innovation continues to offer breakthroughs for us to help our environment. Passive houses are "a standard for houses that are very environmentally friendly (= not harmful to the environment) because they can be kept at a comfortable temperature using no or almost no energy, or a house that meets this standard"[110] This type of house offers natural light through the use of large windows and open stairways, solar chimneys, building with sustainable materials, waste-less layout, and a "green" (roof garden) system[111]. Among some of the advantages of this house design are health benefits, less overheating, thermal comfort, a decrease in the release of harmful emissions, less dust, no or lower air pollution, no dampness or unsavory smells, not as expensive to maintain[112]. Naturally, there are some disadvantages too. These houses have demanding requirements and can cost 10 to 30% more than a traditional type of house[113].

3. Social

We are hardwired for social connection and community. From infancy through to adulthood, our social needs may change, but all in all, we find we need other human beings to survive. The way we interact in a society has an impact on our environment and our quality of life. As much as independence has its benefits; interdependence offers us positive outcomes too. There are numerous sub-topics we could take a closer look at but here I would like us to zero in on sharing, incentivizing smaller family sizes, our work, and machines. Let's begin by looking at the concept of sharing as an element that can help us move towards a happier climate.

3.1. Sharing

At one point or another, you and I had to learn how to share. As you likely know, "sharing" is *"having or using something at the same time as someone else"*[114]. Sharing is one aspect of the growing and global sharing economy, and a positive influencer towards a more eco-friendly future for us all. A sharing economy is *"a peer-to-peer (P2P) economic model focused on acquiring, providing, or sharing access to goods and services. The process is as old as civilization itself but in modern times it is experiencing a revival with the support of community-based online platforms"*[115]. The list of what can be shared is endless (including cars, rooms or whole buildings), and its owners benefit from sharing them at a fee when not in use[116].

For consumers, the shared economy typically offers more financially attractive alternatives. Services like Airbnb are cheaper than traditional hotels by 30–60%[117]; they make access to local life easier, promote the growth of social circles and give visitors a more homely experience, which some visitors desire[118].

Not all sharing needs to be a business transaction. A free version of the sharing economy is growing with people around the world, building connections—and in the case of accommodation—finding locals to accommodate and share accommodation or simply meet. This approach is growing as more people learn of the options that exist. When I analyze it, it seems to hinge on the concept of social proof, which is the psychological influence that persuades us to think or behave a certain way because of the behavior of others[119]. We see social proof at work every day on any social media network through various types of social interactions. Elements such as testimonials, reviews, and comments can give you an idea of someone's degree of social proof (real and perceived). Take a moment to consider new places you've stayed at. To arrive at your decision, did you go online first and read reviews about the establishment before you settled on your choice?

On the topic of accommodation and global networking, two good examples of free hosting communities are the Facebook groups Host a Sister[120] and Couch Surfing[121]. Both these communities are free and connect people, with Host a Sister being only for females, with an emphasis on supporting and connecting with a global sisterhood. To add value members adopt a give-and-take approach to engagement. Someone who is looking for a host will have a lot more luck finding what they're searching for if they have several reviews from people that they hosted successfully. They also use social proof to help them gain credibility, which is akin to a currency that opens more opportunities to be hosted.

Staying in a virtual stranger's home may be unsettling for some. When it comes to sharing, hostels are a viable alternative as another climate-friendly option in comparison to hotels. One study determined that hostels were 75% less carbon intense than hotels with a hotel bed in a hotel averaging 1.18 tCO₂e, and a hostel 0.30 tCO₂e in the years 2019 to 2021[122]. With such a significant difference in terms of environmental impact, the general price difference (with hostels being cheaper) makes this a no-brainer.

Home Away from Home

"Wherever you go, go with all your heart"
- Confucius[123]

We have stayed in hostels in the past but not used communities like CouchSurfing, or Host a Sister. I can certainly understand the appeal, relevance and importance of such communities for people across the socio-economic scale. In our case, we prefer to stay with family and friends instead of a hotel when we are in a city or town that our work or other opportunities have brought us to. We have also opened our home to our family and friends.

This allows us to enjoy two main benefits: lower our carbon footprint since hotels are more carbon-intensive, and spend quality time with people we care about, especially if distance is usually a limiting factor. Staying at a family member's home or that of a friend is perhaps not a novel concept. Throughout my life, and particularly in college, I had friends from all over the world and found this cultural preference prevalent among a number of my friends.

Ulrike

It's not just about sharing vehicles and accommodation to lower your carbon footprint. With some thought, simply cutting down on purchases that require manufacturing can also help you lower your carbon footprint further.

To the Next Generation

"Imagine a world where things are fixed one more time than they are broken."
Laura Kampf[124]

Michael and I both come from humble beginnings. We were fortunate to have been given the opportunities and examples that helped us arrive where we are in life today. When we look back on our childhoods, and the times we grew up in, we realize some of the practices we were taught are worth keeping alive, particularly for the environment.

One particular example that I think illustrates this point well, is from my family. As the children grew up (zero to seven years old), clothes that no longer fit them would be put in a collection and kept. This collection of clothes was kept at one designated family member's home. If anyone had a new baby or their child outgrew their clothing, their parents would search in this collection for clothes they could take for their child that would fit.

Similar to the collection for children's clothes, they also had a collection for toys for extended family. Handling clothes and toys in this manner led to toys and clothes of good quality being in the family for at least two generations.

As we consider our carbon footprints, we see that this is the way to go for our family. As a tradition, we would like to help our children keep this going even for the next generations to come.

- Ulrike

It's reasonable to expect the sharing economy to continue growing as more users join existing and emerging online sharing services.

One major disadvantage of sharing that is worth finding ways to mitigate is any inconveniences that you might face because of choosing to share an item or service instead of owning your very own. Accepting a certain level of inconvenience and finding alternative routes towards what we want to achieve may help us adapt and make environmentally friendly choices for ourselves and our families.

3.2. Incentivizing a Smaller Family Size (Human Population)

Talking about the human population can be a tricky subject. Some hold very strong and polarizing views that I do not. I do know though that the human population is impacting our climate and will continue to do so, especially if we continue living the way we do without making environmental changes. According to the WorldoMeter, the human global population stands at 8.1 Billion and is projected to reach 9 Billion by 2037[125].

The global average fertility rate is 2.3 children per woman, half of what it was 50 years ago[126]. Back then, 4.5–7 children per woman was typical but sadly, many children died at a young age causing population growth to stay on the lower side[127]. Whilst global campaigns to push for birth control are vital, approaching the population challenge from the following perspective may be far more impactful: creating exceptional healthcare systems, and retirement plans to give people a sense of basic social security, encouraging longer lifespans, and therefore causing parents to scale down on the number of children they bring into the world. The more children they can see growing up, the less likely that these parents will have a large number of children to begin with. If many children die in infancy, parents typically have more kids to ensure the probability that they may have survivors.

Back in Time

> *"Study the past if you would define the future."*
> *— Confucius*[128]

In Germany, when we were growing up, large families were the norm. As the data stated, 40–50 years ago, on average, people had 4–7 children, globally.

My grandmother had 11 siblings. This would fall outside of current norms.

Over the past decades, Germany heavily invested in improving the public healthcare system, retirement plans and other features towards creating social security for its citizens. It looks like the nation's efforts paid off.

When Michael and I were younger, we already knew we wanted only two kids. Perhaps if times hadn't changed, we would have wanted more children. That being said, naturally, the carbon footprint of a smaller family is typically less than that of a large family. Though the environment was not a concern for us years ago when we were making our life choices as a family, we are glad that the challenge we are addressing in doing our bit isn't more extensive.

- Ulrike

3.3. Work

Can you think of any other global event in recent times that changed the way we work other than the COVID-19 pandemic? Well, neither can I. As the ancient Greek philosopher Heraclitus once said, "The only constant in life is change."[129] Not only did many of us find ourselves ladled with more time on our hands, but our beautiful planet finally got the break it needed from carbon emissions because of reduced production. In many cities around the world finally had better air quality and lower water pollution[130]. This is even though the temporarily decreased emissions did not translate into lowered atmospheric carbon dioxide:[131]

> *"GDP growth has a positive relationship with increasing environmental pollution and increasing carbon dioxide emissions (Table 4). The elasticity of GDP is 0.02 with a positive sign. This elasticity shows that if GDP increases one percent, CO2 emissions increase by 0.02 percent, assuming other factors are constant (Table 5). With the growth of GDP, production has increased in many parts of the country and the increase in production is accompanied by increased use of natural resources, energy, and various fuels and leading to increased emissions of various environmental pollutants such as CO2 emissions [48]. These results indicate more pollution in the countries with richer economies. As mentioned above, due to the direct relationship between GDP and environmental pollution, carbon dioxide emissions decreased during the COVID-19 outbreak due to reduced production."*[132]

We all saw how organizations and individuals alike had to adapt to survive through the pandemic. Whilst some jobs may never become remote, like those of emergency service workers, many outside of that bracket did or offered work-from-home days for employees. This translated into decreased carbon emissions from less traffic. In industries and fields where remote work was never an option, we were forced to adapt the way we worked so that we could continue being productive, albeit often at just a fraction of our previous capabilities. That fraction is still better than nothing at all. It is therefore reasonable to expect some level of continued change in the world of work and its impact on our environment. The work we do and our activities that come as a result of our earnings impact the environment around us. The International Labour Organization (ILO) shares their insights on this particular context (economic activities and their link to the natural environment) using the following figure:

Figure 1. The economy as a subsystem of the global ecosystem

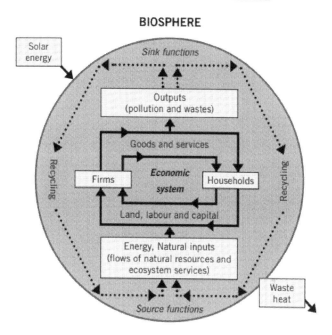

Source: Adapted from Harris and Roach, 2017.

Source: International Labour Organization, 2018[134]

The International Labour Organization (ILO) states the following about the above figure:

"The economic system exchanges goods and services, land, labour and capital between firms and households. Importantly, it draws energy and natural inputs from the biosphere and releases pollution and waste into the ecosystem. The economic system is an open system; it exchanges energy and resources with the global ecosystem within which it is placed. The global ecosystem provides energy and resources to the economy (source functions), and absorbs, stores or recycles the energy and waste produced by the economy (sink functions). The global ecosystem has solar energy as an input and waste heat as an output; other than that, it is a closed system. In current models of economic growth, as the economic system grows within the global ecosystem it requires more resources and energy and generates more waste, making it more difficult for the global ecosystem to perform its source and sink functions. In parallel, some activities within the economic system affects the ecosystem's ability to perform its source and sink functions, both positively (e.g. technology) or negatively (e.g. pollution or destruction of ecosystems). The fixed size and closed nature of the planetary ecosystem imposes a limit to the resources and energy that can be sourced from the ecosystem and also imposes a limit to the amount of waste it can absorb, store or process. In sum, the economy cannot expand beyond the confines of ecological limits."[134]

The ILO concludes by highlighting that if the global community continues to harm the environment, it will affect the way we work in various ways. These ways include increased air pollution and heat stress making working conditions challenging for many, and loss of livelihood for millions of workers globally who work the land (and harvest fish) and depend on its resources[135]. The organization further added that a movement similar to the Industrial Revolution would need to happen to move us from the current way of doing things to using the right resources with a focused incentivized sustainable approach[136]. By this, they mean swapping out ecologically devastating agricultural practices for sustainable agroecological alternatives, switching to renewable energy for fuel at an industry level, and changing the way we work in all spheres.

4. Technology and Innovation

We've looked at CO_2 of living, housing and social aspects that relate to climate change and the future of our co-existence with

Earth. Now, let's look at technology and innovation. From ever-advancing artificial technology applications and capabilities to innovation in the way we construct buildings or sustainable power sources, we live in exciting times. Science, technology, engineering and technology are evolving at break-neck speeds. Here, we will explore some of the aspects of technology and innovation that relate to climate change such as machines and energy, as well as the construction and design of cities. Let's begin with machines.

4.1. Machines

Manufacturing and industrialization put a heavy carbon burden on the Earth. As mentioned earlier in this book, sometimes fossil fuels are used directly in manufacturing and industry and account for 75% of global GHGs and 90% of carbon dioxide[137]. Other times these non-renewable types of energy are used to produce electricity, which is used to power more machines commercially and residentially. In the manufacturing process of many products, some already mentioned materials include cement, electronics, plastics, clothes, iron and steel[138.] We will talk more about machines from a broader perspective in the next section: Technology and Innovation.

One study indicates that machines and equipment use up three times more energy than those of 50 years ago, when we did not have much of the modern machinery that makes life easier and more convenient. Here we are in the information age and modern machines abound, both in our homes and at our workplaces. Like you, I would not complain about the benefits that these machines offer us. After all, who would like to do complex mathematical equations that could take days or weeks to crack, when a machine could do it in a fraction of a second? Who would like to grind corn by mortar and pestle when you can place some grain in a grinder and be done with the task in a matter of minutes? Whether it is a grinder, television, printer, oil press or whatever type of machine, they each give off energy as they work. This energy that machines give off is a little more than 10% of the human-created proportion of the energy from the GHGs[139]. Experts state that locally the impact of increased energy from machines can be felt: *"but locally, energy consumption accounted for by machines and lighting can be much higher and give rise to large local temperature increases, as well as altered rainfall and winds."*[140] One study showed that some specific areas were particularly at risk because of their high energy usage[141].

At around 2010, machines that manage and store our data released enormous carbon emissions that only grew when we entered the data age, leading to a global carbon emission greater than that of Australia's annual carbon emissions[142]. Between 2013 and 2019, the number of data centers from companies such as Google, Microsoft and Facebook tripled to over 500, each typically with numerous rows of servers, supporting extensive computing power, generating heat that the systems work to cool down[143].

4.2. Energy

Let's now take a look at energy from a broader perspective, beyond transport. There exist both renewable energy sources and non-renewable energy sources. Non-renewable energy sources are sources of energy that will run out. Most fossil fuels fall under this category and include coal, petroleum, and natural gas[144].

Here's a diagram presenting the annual global energy potential from both renewable and nonrenewable sources:

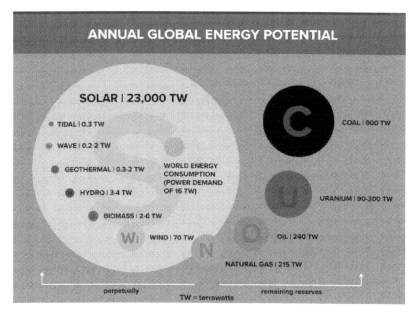

Source: UC Davis, 2021[145]

As renewable energy sources are compatible with creating a healthier planet, we'll focus on those sources next.

4.2.1. Renewable Energy Sources

Renewable energy sources can be replenished; they include hydropower, solar, wind, tidal, geothermal, biomass, and nuclear fusion[146]. Renewable energy sources are considered sustainable and often classified as green. For that reason, we will only take a quick look at each renewable energy source listed above.

4.2.1.1. Biomass

We can harness energy from the dead matter of plants and animals and have been doing so for thousands of years. Renewable energy is referred to as biomass[147]. This type of fuel is often used for heating and the generation of electricity.

Wood, food crops, animal waste, human waste, garden waste, hemp, municipal waste, and landfill gas are some examples of biomass[148]. There is some debate as to whether using biomass is truly good for the environment because of the CO2 that is released when the fuel is burnt:

> *"Burning either fossil fuels or biomass releases carbon dioxide (CO2), a greenhouse gas. However, the plants that are the source of biomass for energy capture almost the same amount of CO2 through photosynthesis while growing as is released when biomass is burned, which can make biomass a carbon-neutral energy source." EIA, 2022*[149]

Though the above may be true, the challenge is that increased demand would spur an excessive harvest of wood and other matter to produce biomass. This could lead to problems like deforestation and decarbonization as a result of lack of crop diversification[150].

4.2.1.2. Solar Power

This is the energy we can capture from sun rays that can then be converted into thermal energy or electricity[151]. Experts state that the amount of energy that the Earth absorbs from solar rays in one day is enough to power the whole world for a full year:

> *"About 30 percent of incoming solar radiation is reflected out into space and plays no role in Earth's climate system. Of the remaining 70 percent, 23 percent of incoming solar radiation is absorbed in the atmosphere, either by water vapor, atmospheric particles, dust and ozone. The remaining 47 percent passes through the atmosphere and is absorbed in Earth's land and sea — which makes up nearly 71 percent of the entire world… The 70 percent of solar energy the Earth absorbs per year equals roughly 3.85 million exajoules. In other words, the amount of solar energy hitting the Earth in*

one hour is more than enough to power the world for one year."
UC Davis[152]

4.2.1.3. Hydropower

Do you know what the world's greatest source of renewable electricity generation is? Hydropower. And it's projected to remain so at least up to the 2030s[153]. This is the power generated from the natural flow of moving water using an elevation difference[154].

Advances in hydropower systems have meant greater conversion efficiency, economic feasibility, lower operational costs, and higher optimization of associated systems[155]. Experts anticipate that wind and solar energy systems will overtake hydropower[156].

Hydropower facilities may include huge dams, modest water sources like that of municipal water facilities, irrigation systems or damless systems using up diversions or run-off water supply[157].

4.2.1.4. Tidal Energy

Another type of power generation using water is tidal energy. This is the type of power harnessed from *"the natural rise and fall of tides caused by the gravitational interaction between Earth, the sun, and the moon. Tidal currents with sufficient energy for harvesting occur when water passes through a constriction, causing the water to move faster."*[158] In a nutshell, tidal energy is a type of hydro-energy specifically generated from waves and tides[159].

If you haven't heard of this type of energy, that's quite understandable. Only approximately 1.5% of the world's total installed electricity is tidal or wave energy, and that translates to 4.5% of global renewable energy capability[160].

4.2.1.5. Geothermal Energy

The core of the Earth, found at 2,900km below the Earth's surface,[161] with a radius of approximately 758 miles[162], is hot, very hot. Temperatures for the inner core measure up to 5,200 degrees Celsius (9,392 degrees Fahrenheit)[163]. Geothermal energy is energy harnessed from this heat emanating from the Earth[164]. Deep geothermal energy requires accessing the Earth's hot temperatures at approximately 2-to-3 miles down[165]. Using enhanced geothermal systems, water is injected into the hot rock underground which either is used directly or for the generation of electricity from the hot water or steam that's been generated[166].

Areas where there are sedimentary basins or volcanoes are typically prime for accessing geothermal energy[167]. Around 2021, geothermal energy was only 0.5% of renewable energy and was growing at a rate of 3.5% yearly reaching 15.96 gigawatts of electricity[168].

Geothermal energy is a more ideal renewable energy as it is also carbon-free and considered to be a key element in the global campaign to decarbonize[169]. Unlike several energy sources, geothermal presents the added advantage of being everywhere:

> " 'Wherever we are on the surface of the planet, and certainly the continental U.S., if we drill deep enough we can get to high enough temperatures that would work like the Boise system," said Jefferson Tester, a professor of sustainable energy systems at Cornell University and a leading expert on geothermal energy. 'It's not a question of whether it's there—it is and it's significant—it's a question of getting it out of the ground economically.' " Yale Environment. 2020.[170]

4.2.1.6. Wind Energy

Wind energy is another renewable and green energy option. This energy is harnessed via wind turbines; the blades are propelled by air—kinetic energy—which is then converted to electricity[171].

Wind turbines can be both on land and offshore, with floating turbines being secured to the sea's floor[172]. European countries and China are forerunners in the space of wind energy generation[173]. In 2022, wind energy accounted for 7.33% of electricity generation globally[174].

4.2.1.7. Hydrogen Energy

Unlike fuels that emit carbon, hydrogen is clean burning and produces water as its only byproduct[175]. Technically speaking though, hydrogen is not an energy source, but rather an energy carrier and is very versatile[176]. Currently, hydrogen is utilized in fuel cells for powering vehicles (particularly heavy-duty trucks) and electricity generation, rocket fuel, and manufacturing and industrial processes[177].

In 2022, global hydrogen demand was at 95 Mt - signifying a 3% annual increase in usage across all major consuming regions, excluding Europe[178]. The latest findings on the importance of hydrogen are encouraging:

> *"The key pillars of decarbonising the global energy system are energy efficiency, behavioural change, electrification, renewables, hydrogen and hydrogen-based fuels, and CCUS (Carbon Capture, Utilisation and Storage). The importance of hydrogen in the Net zero Emissions Scenario is reflected in its increasing share in cumulative emission reductions. Strong hydrogen demand growth and the adoption of cleaner technologies for its production thus enable hydrogen and hydrogen based fuels to play a significant contribution in the*

Net Zero Emissions Scenario to decarbonise sectors where emissions are hard to abate, such as heavy industry and long distance transport." International Energy Agency.[179]

4.3. Construction

The contribution of CO2 that we need to consider isn't just our carbon footprint when we talk about this demand for housing. It's also about the CO2 contribution that we make for building our new homes, especially in the case of homes that are not eco-friendly.

Constructing new homes costs us and the environment significantly. One study carried out in Scotland found that the construction of a two-story, two-bedroomed, two-bathroom cottage added 80 tonnes of CO2e into the environment[180].

In many parts of the world, especially those in the West, wood (and other wood products) were traditionally the fundamental part of construction, mainly due to its versatility, reliability, ease of use in comparison to other materials, and availability/accessibility[181]. Many parts of the world today still use wood, other plant-based products (such as plywood, bamboo joints, thatch, grass, sisal stems, sisal fiber, coir waste, reeds, and straw), and rammed earth to construct their houses[182]. Other materials commonly used in construction are sand, gravel, brickwork, lime, mineral wool, glass, steel zinc, copper, and aluminium[183]. However, cement is the world's second-most-consumed material (after water) and therefore one of the world's preferred fundamental substances for construction[184]. When the right proportions of cement paste and filler are mixed, these substances bind together like glue and help form the structures in the built environment around which includes houses, other buildings, roads, bridges, and dams[185]. During the construction process, it is ideal that the concrete (a substance made from mixing cement, water, sand, and rock[186]) is not allowed to dry out for the first 10 days, a period in which the substance gains strength. By the end of the first 28 days, engineers state that the concrete had gained 90% of its strength[187].

With all the benefits of cement, its price and the environmental impact it has is noteworthy. The limestone used to make cement requires high heat, often fueled by non-renewable energy sources[188]. In addition, the reactions that happen to the chemicals release carbon dioxide as a by-product[189]. As a result, the cement industry emits more carbon dioxide than the aviation industry[190]. With the global housing boom, it is reasonable to expect the amount of CO2 emissions from cement manufacturing to continue to rise unless we make a deliberate decision to choose climate-friendly and sustainable alternatives.

The materials you or building professionals select for construction will influence the homeowner's finances (original cost, and maintenance), durability, and aesthetics [191]. Several factors need to be considered when we choose new materials for the building, including type and functionality of the building, economic aspects (original costs and maintenance costs), availability of materials in your location, availability of specialized and required labor, quality of desired construction materials, mobility costs for the construction project, options for compatible materials required, and personal preference/cultural norms[192].

There is also another factor to consider: the safety of a construction material, for you and/or the environment. Some construction materials are harmful to the environment because of the pollutants they release, the destruction of natural habitats for animals and plants, and the diminishing of natural resources[193]. Others can be harmful to building professionals and people utilizing those buildings because they may be toxic or contain hazardous substances that affect people[194]. Construction materials that may contaminate the environment (through the processes that involve their extraction, transportation and production) include concrete, steel, asbestos, some paints and varnishes, lead, and mercury[195].

Materials used in construction that are not as harmful to the environment as others are known as EP (environmentally preferable) materials[196]. Ideal EP materials have a relatively lower cost to alternatives and a relatively low lifecycle cost; they are energy efficient, locally manufactured from locally derived materials, require comparatively less material to use, are non-toxic, are produced from recycled substances and or salvaged, and are made from materials that quickly replenish[197].

Here's a helpful guide to give you an idea of what this may mean:

Material	Material cost	Life cycle cost impact	Energy efficiency	Water efficiency	Material reduction	Locally manufactured	Locally derived raw materials	Non-toxic	Recycled content	Rapidly renewable	Certified wood	Salvaged
Ceiling tiles	=+	−							●			
Carpet	=	=		●				●	●			
Fabrics (wall/furniture)	=+	=−						●	●			
Resilient flooring	=+	=−						●	●	●		
Interior/exterior paints	=	=						●	●			
Sealants and adhesives	=	=						●				
Steel	=	=			●				●			
Cement/concrete	=	=	●		●	●	●		●			
Insulation	=	−	●			●	○	●	●			
Bathroom partitions	=	=							●			
Wood products	=+	=			●			●		●	●	
Gypsum wallboard	=	=				●	●		●			●
Furniture	=+	=							●	●	●	
Brick/CMU	=	=				●	●					
Roofing	=	=	●						●			
Windows	+	−	●									
Doors	=+	−	●						●		●	
Ceramic tile	=	=						●		●	●	○
Insulating concrete forms	+	−	●						●			
Structural insulated panels	+	−	●						●			
Aerated autoclave concrete	+	−							●			
Exterior finishes						●	○					
Permeable paving	+	−		●					●			

○ Potentially applicable Material & Resource issue, research ongoing
● Applicable Material & Resource issue
(=) Equivalent, (−) Generally less expensive, or (+) Generally more expensive

Image Source: Los Alamos National Laboratory [198]

The fact that we need more accommodation for our growing population and because of the evolving housing trends towards smaller household sizes is certain. We have the opportunity to adapt the homes we already have and those that we intend to build towards accommodation that's climate-friendly.

4.4. Sustainable Materials for Living

Sustainable materials such as bamboo can also be used for other purposes beyond construction. Bamboo is also used to make furniture, paper, clothing, accessories, eating utensils, and even diapers[199]. It is also used in medicine and in environmental preservation methods to deter erosion and deforestation[200].

Many innovative companies are using sustainable materials to replace those that are damaging to our environment. Bambooder is an example of such a company making great strides in innovation and using bamboo and bamboo-derived materials to make extensive products, and reasonably compete with other materials such as steel, aluminum, flax, hemp, and fibreglass[201].

What's more, they ensure there is minimum wastage by using all material from the bamboo. Some varieties of bamboo are edible; the shoots can be cooked and consumed again[202].

4.5. Design of Cities

The United Nations Sustainable Development Goal (SDG) number 11 is to "make cities and human settlements inclusive, safe, resilient and sustainable"[203]. Global stakeholders understand that a discussion about living in the future cannot be complete without discussing our cities, their design, and what we need to do to have a better future as we continue to urbanize. A UN 2023 report showed that we are off-track when it comes to realizing SDG 11[204]. The report highlighted the urgency to rectify this failure.

This cannot be ignored because research shows that 75% of carbon emissions are from urban areas worldwide, largely due to our transportation activities and our buildings[205]. We will always have cities, so the question is: How do we make cities climate-friendly? Climate Foresight, the Centro Euro-Mediterraneo sui Cambiamenti Climatici (CMCC) observatory on climate policies and futures suggests the following:

> *"Inverting this trend will rely on policy frameworks, technological advancements, and the adoption of sustainable practices that can redefine the way we think about the buildings in which we live, work and play. This includes rethinking the way we certify the sustainability of buildings. In fact, there is a growing recognition of the deep interconnection between decarbonization in the building sector and our ability to address climate change leading to increasing emphasis on energy-efficient building design, renewable energy integration, and sustainable building materials… Governments and organizations worldwide are implementing policies and regulations to promote energy efficiency and low-carbon buildings." Foresight, 2023*[206].

Dark Matter Labs, in their project "Desire" make excellent points about what designing for a climate-friendly future requires. Climate-friendly cities will require changes in typography, our concept of comfort, material innovation, radical sharing, extended ecological services, care, collective intelligence, and spatial computation[207]. Note Dark Matter Labs's illustration of the same below:

Source: Dark Matter Labs, 2023[208]

We need a fundamental transformation at speed and scale; we need to build capacity for innovation across sectors, and be more collaborative in our collective approaches[209]. Dark Matter Labs further adds in their paper, "Designing Our Future" (Invitation Paper v.01), several suggestions on pathways which I support and have placed as recommendations for the end of this chapter[210].

4.6. Artificial Intelligence and Other Technological Advances

Artificial intelligence, or AI, is "the science and engineering of making intelligent machines"[211]. AI may be gaining prominence in the mainstream but it was born in the 1950s[212], albeit not as advanced or with as many real-world applications as we now see today. If you are like me, you've probably played around with ChatGPT just to see what all the fuss was about. You may also have found yourself interacting with a bot more times than you would like and sometimes, maybe you didn't even know you weren't communicating with a real person. These activities and automation of administrative tasks, intelligent feedback and assessment, collaborative learning and virtual classroom, adaptive learning paths, virtual reality and augmented reality, intelligent content creation, predictive analytics decision making and bias detection are just some of the many use cases for AI[213].

Now, you may be wondering: what does AI have to do with climate change and adequately addressing the problem? The answer to that lies in the increasing capabilities and applications of AI that we can harness today towards fighting climate change; some are already utilized at present and the are innovations that we can expect to materialize soon. Here are some projects already underway, according to the UN Climate Change website (AI for Climate Action: Technology Mechanism supports transformational climate solutions)[214]:

- Innovative adaptation technologies: with immense computational power and smart algorithms, systems can be programmed to alert people in disaster-prone areas of pending catastrophes allowing them to take necessary precautions.
- Agri-food systems and crop management optimization processes: agribusiness has many variables, with AI, agri-businesses can make smarter decisions with the variables they have, even as they change. AI can also lower water wastage, encourage more sustainable practices and increase food production.
- Renewable energy systems: Some of the limitations of renewable energy solutions that are available today lie in their efficiency, reliability and cost. Renewable energy

systems powered by AI can predict demand, and can assist with seamless grid integration and optimization[215].

Other than AI, other technological advances can help us fight climate change: carbon capture, removal, and storage; renewable energy, batteries and energy storage; smart homes, buildings, cities, grids, agriculture; and remote sensing of greenhouse gas emissions[216]. Fortunately, we live in an age when advances are emerging at incredible speed. I believe the more investments that are made available to spur these on, the more hope we have of using technology as an additional tool in our arsenal against climate change.

4.7. Energy of the Future

Though renewable energy has many benefits for us and is a far better option than fossil fuels that release carbon into our atmosphere, there is room for further innovation and technological advancement. Indeed, we must address the shortfalls of green energy that threaten to negate the gains that this green energy offers. Here is what we are contending with:

> *"Solar and wind facilities require up to 15 times more concrete, 90 times more aluminium, and 50 times more iron, copper, and glass than fossil fuels or nuclear energy. Green technologies require the use of rare minerals whose mining is anything but clean, causing heavy metal discharges, acid rain, and contaminated water sources. It is estimated that three billion tons of mined metals and minerals will be needed to power the energy transition. A recent report identifies the mining industry as the second-most-polluting industry in the world- and the green energy transition would catapult this to number one, meaning our planned green transition is far from being green. Global mineral reserves supply chains and*

> *reserves may be insufficient to supply enough metals to manufacture the necessary non-fossil fuel industrial systems, meaning we don't have enough access to materials for the planned green transition." Dark Matter Labs, 2023*[217]

Some argue that the lag in advancement in better green energy solutions is due to the influence of incumbent stakeholders in the industry who have interests and investments embedded in the status quo of energy solutions. This is easy to believe from an empirical perspective. I have lived in Western countries with excellent levels of sunshine almost year-round. However, when you look at the number of residents in the community using solar energy to power their homes, you might be surprised by how few utilize solar power.

4.8. Carbon Neutral

As we discuss how we live, we need to look at carbon neutrality, which refers to creating "a balance between emitting carbon and absorbing carbon from the atmosphere in carbon sinks"[218]. A carbon sink is anything that captures carbon dioxide and includes natural carbon sinks like the ocean, soil and forests[219]. These are important because not only do they reduce carbon in the environment but they typically absorb more carbon than they release[220].

5. Recommendation

The growing global population, gigantic housing demand, our global energy demands, the impact of our machines on the environment and so much more that we've discussed should help us understand the need for change in these aspects of climate change too. For that reason, I hope you can find the following recommendations fitting:

5.1. Evolve Our Value Systems and Economy

Living in the future will need us to change, beginning with our value systems. The insights shared by Dark Matter Labs' project, Desire, present many salient points on this topic. You may refer to a seminal quote earlier in this Chapter, in the section "Co2 of Living" about our need to change our concept of comfort.[221]

Allowing ourselves to change our standards and adjust our comfort levels may also mean passing down toys, clothes and other items to the next generations instead of always purchasing brand-new items when the old ones are still perfectly usable. It may also mean using reclaimed wood instead of cement on your house's extension, instead of purchasing numerous bags of cement or sheets of steel.

5.2. Adopt Renewable (carbon-free/ carbon-low) Energy Solutions

Solar power solutions are available for any type of context and budget. Since price is not a prohibitive factor for many of us who still rely on non-renewable sources, switching to solar power is one of the easiest contributions we can make to lowering our carbon footprint and contributing positively towards handling climate change. Other renewable forms of energy such as wind, biomass and hydrogen are still open to us to consider using. I highlight solar, purely from an accessibility perspective as even those of us from the most challenging financial background may afford solar lights at the very least. It's easy to argue that despite the favorable cost aspect of solar energy solutions, many people who can afford it simply do not. Others with the financial capacity to purchase home solar solutions that replace their grid setup argue that solar panels are ugly. Companies have started selling solar panels that have a more aesthetically pleasing appearance. Solar tiles or shingles out there in the market include Tesla's Solar Roof, and solar shingles by Certain Teed, Suntegra, Forward, and Luma[222].

5.3. Hire Green Builders

When you decide to build, choosing a green builder can be the decision that makes the biggest difference in ensuring you have a more climate-friendly home. With more countries adopting LEED certification, your chances of finding a conversant professional with this certification are growing[223]. Since there are no certified builders and only certified homes, your search should be focused on green builders[224].

5.4. Construct Eco-friendly Homes to Meet Global Demand

With the projected demand for 2 billion homes globally in the next coming decades (over 80 years from 2018)[225], the climate will be affected by the carbon emissions from the actual construction of the houses and the manufacturing and production of construction materials such as cement and steel in particular. We have an array of alternative materials such as composite materials, bamboo, reclaimed wood, recycled materials, and earth that we can consider using for our housing projects.

Construction professionals who are not familiar with working with these types of materials have the opportunity to embrace these alternatives and even educate their clients about their eco-friendly qualities as viable contenders for their next building projects.

5.5. Continue Professional Development (Building Experts)

Cost alone should not be the discriminating factor when we choose a particular building material for a project. With a life cycle assessment of all the materials that will be used, we can make informed and climate-friendly decisions throughout the construction of any building. The Carbon Leadership Forum provides a helpful guide to assessing the environmental impact of a building project[226].

5.6. Construct New Features Using Eco-friendly Materials

Use eco-friendly materials for features you add to your home. A bamboo outdoor recreational area could look beautiful and more cost-effective than a conventional option. Of course, this depends on the availability of the material where your home is. There are several eco-friendly options that you can choose from[227].

5.7. Use More Environmentally Friendly Appliances

Using energy-saving devices like A-rated fridges and freezers amongst other energy-saving electrical or low-carbon emission devices is an important contribution that you and your household can make towards a greener future. Adequate space heating is extremely important. Heat pumps are one device that you can consider[228]. A heat pump is a device that transfers heat from one place to another[229]. Though these devices may not be ideal for places where the temperature goes below freezing, and they require electricity to function, among other disadvantages, they also have several benefits. These benefits include a higher energy efficiency as they pump out cool and warm air[230].

5.8. Go Green for Businesses

In addition to going green at home, our businesses need to go green too. Hotels and other businesses in the accommodation industry are one type of business that could help our fight against carbon emissions if they made efforts towards lowering their carbon emissions contribution.

5.9. Take Your Business to Green Enterprises

We perpetuate what we support. By selectively purchasing goods and services that have a green focus, we ensure their continued financial viability, presence and success. Even simply letting your favorite businesses know that you have a focus on green solutions may prompt them to pay attention and perhaps get on board when they can, simply to remain competitive and not lose your business.

5.10. Plant a Tree

We need to invest in our carbon sinks and contribute towards carbon neutrality. Planting trees is one easy, cheap and convenient way for us to work towards that. Trees, as we know, make a great positive contribution to the environment and the statistics back that up: "According to estimates, natural sinks remove between 9.5 and 11 gigatonnes of CO2 per year. Annual global CO2 emissions reached 37.8 gigatonnes in 2021."[231]

6. Conclusion

As we conclude this chapter, let me offer you one story which may help us think about how we are living and what we may need to realize about our sense of satisfaction. This story is from which the quote under this chapter's heading was sourced:

> *"At a party given by a billionaire on Shelter Island, Kurt Vonnegut informs his pal, Joseph Heller, that their host, a hedge fund manager, had made more money in a single day than Heller had earned from his wildly popular novel Catch-22 over its whole history. Heller responds,* **"Yes, but I have something he will never have ... enough."** *Enough. I was stunned by the simple eloquence of that word—stunned for two reasons: first, because I have been given so much in my own life and, second, because Joseph Heller couldn't have been*

more accurate. For a critical element of our society, including many of the wealthiest and most powerful among us, there seems to be no limit today on what enough entails."[232]

Do you and I know what "enough" is? Can we try to find it for the love of future generations and the one planet we have? In this chapter, we focused on how we live and how we can live in the future if we choose to. In the next chapter, we are going to focus on our food and food systems.

Chapter 3:
FOOD OF THE FUTURE

"In addressing climate change by reducing emissions, we are preventing the worsening of health conditions around the world, and by improving so many different conditions that can be improved through climate measures—such as improving food and water, food security and water safety—we are actually improving health conditions."
– Christiana Figueres, Former Executive Secretary of the United Nations Framework Convention on Climate Change[233]

Did you know that globally, the leading cause of death right now is a poor diet?[234] In the previous chapter, we looked at how we live and what a greener future would demand of us. In this chapter, we will talk about our daily diets—what we eat and our health as it relates to climate change. We will also look at what we can do to ensure that both we—humanity—and the planet realize optimum health.

1. What Is Food?

This may seem like a question not worth asking but What is food? Really? Surely, food is simply what we eat. The dictionary definition of food is the "substance(s) consisting essentially of protein, carbohydrate, fat, and other nutrients used in the body of an organism to sustain growth and vital processes and to furnish energy."[235] If we understand what food is or is meant to be, we may come to the shocking conclusion that some of what we consume and consider food, simply is not. Here are the classifications from the NOVA food system[236]:

- Unprocessed or Minimally Processed Foods: This includes apples, fresh oysters, and unpasteurized fresh milk with no additives.
- Processed Culinary Ingredients: These are substances that we add to our foods that are made through methods such as refining, pressing, grinding, milling, and spray drying. Salt, sugar, and oil fall into this category.
- Processed Food: This is food that has had some relatively simple processing done with certain methods such as tinning/canning, concentrating, fermenting, curing, and smoking. Canned fish, pickles, salted or sugared nuts and seeds, some kinds of butter, and some cheeses go under this category. Traditional diets from around the world often use these methods and are typically connected with good health. We will talk more deeply about diets further in this chapter.
- Ultra-processed food: These are foods manufactured with the cheapest products, designed to have long shelf lives and maintain the longest intellectual property. Typically this category of "food" contains derivatives from corn (starch, oil), soy, rice, and meat. Often, these foods are made from the byproducts of the production of other foods and/or substances that have undergone additional processing.

You may have seen some of the names of these ingredients on the packaging of some supermarket foods: hydrogenated or interesterified oils, hydrolyzed proteins, soy protein isolate, maltodextrin, inverted sugar, and high-fructose corn syrup. Essentially, these are foods that have sensory or cosmetic intensifying additives and can include foods from both the first and third groups if further processing and additives are incorporated. Think of highly processed ice cream, yoghurts, sausages, most store-bought candies, mass-produced breads, and some alcohols such as whiskey and vodka. We will explore ultra-processed foods further later on in the book.

2. What Is Health?

The main connection between our food and climate change that is most relevant for this discussion is health: your health, my health, and the health of the climate. When it comes to our health as mankind, the WHO defines health as *"a state of complete physical, mental and social well-being and not merely the absence of disease or infirmity"*[237]. We probably all know that healthy food plays a vital part in keeping us healthy all around, alongside other lifestyle habits such as good sleep hygiene, and exercise[238].

The majority of food you would consider whole food and many traditionally processed foods are considered healthy for most of us. However, that is not the case with ultra-processed foods. These foods are often very soft and energy-dense[239]. These two qualities make them very easy and quick to consume. Their ingredients are often high in sugar and carbohydrates, making them addictive, and therefore the kind of foods that many people who struggle with weight find hardest to let go of. Poor nutrition—malnutrition—is about not consuming enough food (undernutrition) or enough of the right food[240]. Overweight and obese are two of the subsequent and common causes of prolonged malnutrition.

Thirty nine percent of adults worldwide were classified as overweight or obese in 2016[241]. More worryingly still, some analysts project that global overweight and obesity will hit 51% by 2035[242]. Being obese and especially overweight comes with many negative health risks such as high blood pressure, high cholesterol, diabetes (type 2), coronary heart disease, stroke, gallbladder disease, osteoarthritis, sleep apnea, some cancers, body pains, and some forms of mental health challenges such as depression, and anxiety[243].

Poor nutrition does not only affect us directly. It can also impact unborn generations. We may all know how important a pregnant mother's nutrition is for her and her baby's health. Studies suggest though that in addition, the status of nutrition of both parents can have intergenerational consequences: *"Periods of nutritional restriction during both parents' fetal life can have intergenerational consequences, affecting their offspring's fetal and postnatal growth."*[244] An example from my home of choice—the Netherlands—from the Scientific American[245] illustrates this further:

The Dutch Winter

1944 - 45

"Turning a blind eye to our history has not saved us from its consequences."
— Cicely Tyson, Just as I Am[246]

"The Hongerwinter was a major famine that took place in the Netherlands, particularly in the Nazi-occupied western part of the country. From November 1944 until the liberation of the Netherlands by the Allies on 5 May 1945, 22,000 people died and 4.5 million were affected by the direct and indirect consequences of the famine. The "Dutch Hunger Winter" was caused by a number of reasons: in addition to an exceptionally harsh winter, bad crops, and four years of brutal war, the Nazis imposed an embargo on food transport to the western Netherlands in September 1944 in retaliation for the exiled Dutch government supporting the Allies in liberating southern parts of the Netherlands. The population was forced to live on rations of 400-800 calories per day; to survive, people had to eat grass and tulip bulbs. Besides the aftereffects on the Dutch survivors such as poor physical health, the famine resulted in long-term effects on the descendants of the Hongerwinter generation. Babies born during this period were

conspicuously small and extremely vulnerable to diabetes, schizophrenia, and lung diseases." Environment and Society. (n.d.).[247]

- Ulrike

Through this tragic example, we see the far-reaching consequences of poor nutrition. During this period, these people did not have the means to overcome these dietary challenges under famine and unprecedented oppression. However, the likelihood for you and me today, wherever you are, is that you have an abundance of access to the nutrition you and your family need. The importance of good health, and doing what you can to maintain it—especially with good eating habits—cannot be understated.

3. C02 of Food

As with transportation, food and food systems also contribute significantly to CO2 emissions and other GHG emissions. A food system is *"a complex web of activities involving the production, processing, transport, and consumption. Issues concerning the food system include the governance and economics of food production, its sustainability, the degree to which we waste food, how food production affects the natural environment and the impact of food on individual and population health."*[248].

In 2015, research showed that 34% of greenhouse gases caused by humans were emitted through food systems, with roughly 50% of those GHGs being CO2[249]. Of that 34%, 71% of emissions came from "associated land use and land-use change activities" (LULUC)[250]. Forests are important for protecting the Earth by absorbing significant quantities of CO2 and subsequently helping to protect our climate[251]. Trees in our forests absorb more CO2 than crops (such as corn) that we grow in fields. As a result, the fewer forests we have, the more CO2 remains in the atmosphere[252]; 50% of the land that can be inhabited is presently used for agricultural activities with over 75% of it being utilized for livestock agriculture in particular[253]. The diagram below illustrates the breakdown of global land area for food production:

Source: OurWorldinData.org[254]

Furthermore, if we add up pastoral areas with areas used for growing animal feed, then livestock farming amounts to 77% of all farming land worldwide[255].

The food type that releases the most CO2 and GHGs is meat; it takes up the lion's share (57%)[256]:

GREENHOUSE GASES CONTRIBUTION BY FOOD TYPE IN AVERAGE DIET[3]

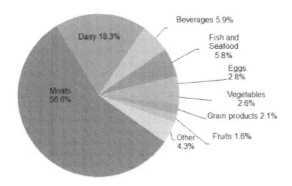

Source: Center for Sustainable Systems - University of Michigan[257]

Since the location where food is grown, processed, and ultimately sold to us, the consumer, are not necessarily the same places, the concept of "food miles"—essentially the distance covered by food items from consumption—becomes an additional factor in the CO2 of food systems. The higher the food miles, the more detrimental it is to our environment and climate. According to research, 19% of GHGs from global food systems are attributed to the transportation of that food[258].

At various parts of the food production process, GHGs are released into the atmosphere. Livestock such as cattle, sheep, and goats have digestive tracts with microbes that help them break down food. During that breakdown process, these microbes cause the release of methane—a greenhouse gas, that these animals subsequently expel from their bodies[259]. When we use fertilizer to grow our crops, the fertilizer is broken down by microbes which leads to nitrous oxide (another environmentally harmful GHGs) being released into the environment[260]. Nitrous oxide (commonly known as laughing gas) not only has the potential to harm our crops directly, but its release adds to the ozone layer and the formation of aerosol (a suspension of fine solid or liquid particles in a gas)[261]. In addition, inhaling this gas can lead to dizziness, unconsciousness, infertility (with long-term exposure), and even death[262].

Other than the impact of fertilizer on our environment, rice cultivation, burning of crop residue, and the use of carbon-based fuels in the production process also contribute towards CO2 of food and food systems[263].

Hopefully, I've painted a clear enough picture of CO2 emissions from food and food systems. Let's now take a closer look at our daily diets.

4. Diets

When we refer to "diet" here, we are talking about it from the perspective of our daily eating habits. Typically, traditional diets from around the world are considered healthy, with most of the ingredients locally grown or sourced, often unrefined and fresh[264]. Such diets include the Mediterranean, Latino, Asian, and vegetarian diets.

Here we will take a moment to focus on just a few of the most prominent diets and how they impact climate change and our health. These diets are plant-based, vegan, vegetarian, Mediterranean, and ultra-processed.

4.1. Plant-based or Vegan Diet

According to the Vegan Society, 1–2% of the world's population eats the vegan diet[265]. It consists of food made from plants[266]. A plant-based diet focuses on consuming foods from plant sources; even though one may eat very few animal products, the majority of nutrition emanates from plant and plant sources.[267] In our case, we eat a whole-food plant-based diet. It avoids all animal products and includes unprocessed or minimally processed foods[268]. The distinction between a plant-based diet and a vegan diet may seem confusing to you at this point but there are some significant differences. One of the biggest differences that is most relevant to our discussion is that a vegan diet does not include the consumption of animal products, whilst a plant-based diet allows for it[269].

4.2. Vegetarian Diet

Here we have some overlap with the definitions of "plant-based" and "vegan". A vegetarian diet "excludes meat, meat-derived foods, and, to different extents, other animal products". This definition would include, among others, ovolactovegetarian (including eggs and milk) and vegan diets"[270].

The Heart Foundation presents these diets and a few others to show the difference:

Different styles of plant-based eating

Name of diet	Description
Semi-vegetarian or flexitarian	includes eggs and dairy may include small amounts of meat, poultry, fish and seafood
Pescatarian	includes eggs, dairy, fish and seafood excludes meat and poultry
Ovo-vegetarian	includes eggs excludes meat, poultry, fish, seafood and dairy
Lacto-vegetarian	includes dairy excludes meat, poultry, fish, seafood and eggs
Vegetarian (a.k.a. lacto-ovo vegetarian)	includes eggs and dairy excludes meat, poultry, fish and seafood
Vegan	excludes all meat, poultry, fish, seafood, eggs and dairy

Credit Source: Heart Foundation[271]

4.3. Mediterranean Diet

Considered one of the healthiest diets, the Mediterranean diet consists of plant-based foods and healthy fats from diverse food sources such as olive oil and fatty fish[272]. With this diet, you can consume what you would with the average plant-based diet, such as whole grains, beans, nuts, fruits, vegetables, and extra virgin olive oil[273].

4.4. Ayurvedic Diet

The Ayurvedic diet is an Indian traditional diet. It is considered healthy emphasizing the use of food for optimum holistic health by personalizing the foods you consume based on your personality[274]. Though it is rooted in religious practices going back many centuries, people around the world, and particularly in the West have awoken to the health benefits of an Ayurvedic diet, such as its gut health-friendliness, potential to reduce chronic disease, and weight loss potential[275]. Some attribute this partially to the global effect of the coronavirus pandemic and the impact that immunity played in recovering from the illness.

The Ayurvedic Institute presents a comprehensive list of foods that people following this diet may consume, including fruits, nuts, grains, animal products, herbs, and some spices such as turmeric, thyme, vanilla, poppy seeds, fennel, ginger, garlic, nutmeg, peppermint, mustard seed, saffron, fenugreek, black pepper, and cinnamon[276].

4.5. Ultra-processed Food Diet

We looked at ultra-processed foods (UPF) at the beginning of this chapter in quite some detail. If you cannot recreate that food in your own kitchen given the same ingredients, then it is likely a UPF[277]. As mentioned, UPF contains various substances, many not found in the typical residential kitchen: *"either food substances never or rarely used in kitchens (such as high-fructose corn syrup, hydrogenated or interesterified oils, and hydrolyzed proteins), or classes of additives designed to make the final product palatable or more appealing (such as flavours, flavour enhancers, colours, emulsifiers, emulsifying salts, sweeteners, thickeners, and anti-foaming, bulking, carbonating, foaming, gelling, and glazing agents)...[278]".* These foods are common in Western diets and have been linked to the growing problem of obesity and overweight in many countries around the world, including the United States, Canada, Brazil, and the United Kingdom[279].

5. The Humane Approach

Have you ever wondered about how we, as human beings, decide which animals are food and which aren't? In some cultures, dogs and cats are pets and considered part of the family. In other countries, these animals are as much food as eggs, fish, chicken, and cows. Travel around the world and you will find that whatever you consider not food, is someone else's delicacy. When we look at the nutritional profile of these animal-based proteins, we can understand why people make these culinary choices, and with them comes great demand. The World Economic Forum shows the figures for 2019 of the animals that are slaughtered for our diets each year[280]:

- 50 Billion chickens - this does not include male chicks and unproductive hens killed in the egg production business.
- 1.5 Billion pigs (for our pork, bacon, and ham production), a number that has tripled over the last 50 years
- Half a billion sheep
- Over 300 million cows
- Approximately 444 million goats
- 150 million tonnes of seafood: The numbers of individual fish are hard to estimate because the industry quantifies them based on collective weight.

Most animal-based proteins are rich in protein and various nutrients that human beings need such as omega-3[281] in fatty fish and Vitamin B12 in meat[282]. The variances in diets are often influenced by culture[283], ecosystems, and socio-economic factors. As we explore, it's necessary to approach the topic with equal respect for each culture and our diversity as people.

At the same time, the discussion of animals as food for human beings must include the topic of being humane in our animal farming or catching methods. Being humane means having compassion, sympathy, or consideration for animals or other people[284].

The details of how animals are farmed or caught for our dietary preferences remain a mystery to most of us. For some of us who do know what goes into getting our steaks, chicken, and pork chops on the table, we may remain somewhat disconnected from the fullness of the processes, and whether these processes are done in a humane manner or not.

Countries around the world have instituted some standards to which farmers must adhere for the well-being of farm animals. In the United Kingdom, for example, they established five freedoms that all farm animals must be given[285]; freedom from fear and distress, freedom to express normal behavior, injury and disease, freedom from pain, freedom from discomfort, and freedom from hunger and thirst. How well farmers work towards ensuring the provision of these freedoms is worth researching. However, through an empirical assessment, still much more can be done.

Since statistically, most of us consume beef, chicken, and fish. Let's take a closer look at what the lives of these animals look like as they are reared or harvested for our consumption.

5.1. Cattle

If you have pets, then I am sure you've noticed the type of personality your furry family member has. Maybe your pet is rambunctious or perhaps shy but whatever the case, you recognize that that behavior is just part of your beloved pet's character. Cows are also unique in their personalities and have their own likes and dislikes[286]. Scientists have even found that cows enjoy listening to slow music as it helps them relax.

Cows are also smart and have social hierarchies, with leadership and even friendships that they nurture among themselves. A mother cow cares a great deal about her calves, so much so that she is inclined to allow them to suckle for up to a year[287]. The emotional impact of being separated from their mothers can linger with a cow even through its adult years[288].

The lives of cattle raised for food differ depending on what the individual animal is intended for by the farmer. Dairy cows are kept for three or four years for milking, a relatively short period due to how taxing the constant milking is on the cow[289]. Just like us, cows only produce milk if they become pregnant. As such, farmers need to ensure a cow gives birth to one calf a year to keep milk production going[290]. Then, when a cow's milk production days are behind her, she's slaughtered and butchered for her meat.

Young female dairy cows are beneficial for farming but male cows are not because they can't lactate. When a male calf is born, it's swiftly taken away from its mother and put into a solitary stall for 16 to 18 weeks, thereafter. Some farmers resort to feeding veal calves only liquid diets to ensure that the meat produced from them is pale and tender[291]. Eventually, it's slaughtered. On our plates, this meat is savored for its tenderness, characteristic of the youthfulness of the young animal.

Cattle bred for beef may have more freedom to roam and eat from pastures if they are on free-range farms. However, whether free-range or not, conventional farming of them includes castration (removal of the testes), dehorning, disbudding (removal of horn buds from young cattle), and branding—all without pain relief[292].

5.2. Chicken

We domesticated chickens 3,500 years ago[293]. Beyond the taste that many of us savor, chicken has so much more to it. Several studies over the years have found that chickens are self-aware; they are capable of demonstrating self-control, have unique personalities, and can do some basic mathematics (addition and subtraction[294].

Another study found results implying that chickens are capable of reasoning, logical inference (deriving meaning or logical conclusions from something else), self-assessment, communication with each other, social cognition and complexity, telling individuals apart, perspective-taking, social manipulation, emotions (both negative and positive emotions), and even social learning[295].

Chickens are farmed in differing ways around the world. In commercial production systems, it's common for chickens to be reared in enclosures with close control of the lighting, ventilation, feed, and water[296]. These are typically environments artificial to their natural environments thousands of years before we began to breed them for human consumption. For economic efficiency, these birds have limited space to live in during their lifespan; typically this means one square meter accommodating 19 two-kilogram chickens[297].

Our breeding of chickens over the centuries for maximum production value has led to poultry prone to developing serious health problems during the production process such as leg problems in broilers (chickens raised for their meat), heart attacks, and also skin diseases[298]. Apart from the confinement—often in cages—some of the inhumane methods used by some farmers include immediate separation from their mothers, debeaking or beak trimming (without anesthesia), and starvation (to encourage egg laying in a process called molting)[299].

As a result of confining animals in such cramped conditions, chickens and other poultry confined in such ways are also prone to other diseases; farmers use antibiotics to treat them. These antibiotics don't just go away once the animal is treated. When a consumer eats these chickens, the antibiotics enter the digestive system and are subsequently absorbed into the consumer's bloodstream, and further circulated throughout the whole body through the blood[300]. A diet high in poultry and other animals that receive heavy doses of antibiotics is believed to have contributed to significant levels of antimicrobial resistance[301].

Statistics project that by the year 2050, this mass resistance to antibiotics will lead to approximately 10 million people dying globally from common infections simply because antibiotics will not be effective for them anymore[302]. Furthermore, experts state that the magnitude of this problem, at that point, will exceed global deaths due to cancer[303]. Essentially, this implies that we are setting ourselves up for medical challenges as we consume chicken and other animals, and simultaneously medically setting ourselves back decades—partially because of our dietary preferences.

5.3. Fish

I used to think fish weren't the smartest of creatures. Perhaps you think that now. Interestingly, researchers have found that fish don't only have intellectual capacity but maybe they are just as smart as other vertebrates, or even more so[304].

Not only have researchers found that fish can learn from each other, evade traps and retain the memory of those traps for up to a year, know their fish social hierarchies, and retain memories of complex maps of their whereabouts, but they can also pass down information to their offspring, including migration routes[305].

Factory farming is one way in which fish end up in the stores and on our plates. Fish factory farming, also known as intensive fish farming, is where fish are kept in containers or nets, in cramped conditions for the majority of their lives[306].

Due to the rapid buildup of fish feces in these containers and the stress of living in such cramped and unnatural environments, these fish sometimes have suppressed immune systems, making them susceptible to illnesses[307]. When these fish get sick, as with chicken and any other animal raised for food, they are administered antibiotics to help them regain health. As I have already stated, regarding chicken, these antibiotics have a massively negative impact on our health when we consume too many animal products from animals that receive significant amounts of antibiotics. In the UK as far back as 2016, some stakeholders warned the government that because 80% of all antibiotics used in the US and 45% in the UK are given to farm animals, there was a genuine risk of drug-resistant bacteria developing[308].

At this point, I would like to share with you a more personal experience that shaped my present-day dietary choices:

That One Chicken

"There are two things nobody should ever have to watch being made, sausage and laws."
~ Mark Twain[309]

In my case, it wasn't laws or sausage that I watched being made. It was chicken.

My family and I mainly eat a plant-based diet. My wife and daughter led the way and I followed suit some time thereafter. My son has differing dietary preferences. However, before the rest of us went the route of a plant-based diet, just like most people I enjoyed my steaks, chicken, milk, fish, and eggs. Interestingly enough, some time before I committed to this significantly different dietary lifestyle, one instance had a profound and personal impact on me. This was the day I saw a chicken being slaughtered for me to eat.

I won't go into great detail about it but I will say that the memory of it is seared into my mind. Though this chicken died almost immediately, I was horrified to later learn that a beheaded chicken can survive for longer than you can imagine—without a head! One chicken lived for up to 18 months without a head on a farm in America when it ran away from being someone's food back in 1945[310]. For a year, I could not bring myself to eat any chicken. As I said, at that time, I was still eating an omnivorous diet but this very visual and audible experience put a blight on my tastes. It caused me to think deeply about what eating chicken involved.

Though I resumed eating chicken a year later, I guess this experience planted a seed in my mind about foods made from animals and the reality of what goes into it for that piece of food to end up on my plate.

As I now continue a committed plant-based diet and my family does the same, I have never been in better health. Several months ago I completed my 120th marathon in New York and continue to enjoy a healthy lifestyle with my family. Though the switch was hard many years ago, I am glad I made the change. I never look back.

I sleep better for it—from a conscience sense—and I feel better all-round: so does my family.

- Michael

So, ultimately animal welfare—how we treat animals that we consume as food—is more about economics, than anything else. That is where we see their value, as commodities and things, and not beings. Yes, they are not humans but are they simply just "things"? Science gives us some answers: Studies tell us that animals we consume for food are more cognizant of their lives and experiences than we often think:

> *"Animals are unlike other things that we own because, unlike cars, cell phones, and sofas, the overwhelming portion of animals we use for food, clothing, research and testing, etc. are sentient; that is, they are subjectively aware. They have interests—preferences, wants, and desires. It costs money to protect their interests. And, for the most part, we spend that money to protect their interests to the extent that we get a benefit from doing so."* Francione (2022)[311]

This is a complicated and uncomfortable discussion, without a doubt. At the very least, we can take a deeper look, pause, and think: What makes your dog or cat a friend or family and a rabbit, chicken, or pig "food"? What does eating meat and other animal-based products mean and entail?

6. Food Waste and Food Security

Earlier in this chapter, we talked about the growing problem of obesity the world over and how by 2035, it's expected that 51% of us will be overweight or obese[312]. In the US alone, 58% of adults will be obese by 2035[313]. In 2022, the WHO stated that there were more than 1 billion people with obesity: 340 million being teenagers and 39 million being children[314].

In 2022, according to the Global Report on Food Crises, roughly 258 million people faced food insecurity and even starvation in over 58 countries around the world[315]. This is the highest number in this UN backed annual report's seven year history[316].

With the bleak predictions of global obesity and overweight and the disturbing reality of acute starvation in countries around the world, statistics show that roughly 33% of food meant for us to consume is lost or wasted globally[317]. Furthermore, they state that this food would be enough to feed all the world's people facing food insecurity and starvation with plenty still left over[318]. Thankfully, governments around the world understand the problem and through the United Nations Sustainable Development Goals have committed to reducing food waste by 50% by 2030[319].

The reality right now though is this lost or wasted food ends up contributing even more to our CO2 emissions. WFP makes a poignant point about this: *"If wasted food were a country, it would be the third-largest producer of carbon dioxide in the world, after the USA and China."*[320]

7. How We Grow Food

A conversation about how we grow our food needs to start with understanding the differences between organic and non-organic food. Organic food is grown via methods that do not use synthetic chemicals or biochemically engineered inputs such as genetically modified organisms (GMOs), pesticides, or artificial fertilizers[321].

Non-organic food is therefore food that is grown using GMOs, pesticides, and or fertilizer. Some studies have found an association between pesticide exposure and obesity, leading researchers to conclude that pesticides utilized in growing our food contribute to negative health outcomes such as excessive weight gain (obesity) and hormone imbalances as the chemicals in pesticides alter the way our bodies make, store, and utilize fat[322]. Pesticides are not the only chemicals to worry about in our foods that contribute to unwanted weight gain: so are hormones injected into some of the animals that end up becoming our food and even some of the chemicals that line the packaging of our food[323].

Reports also state that organic food has higher nutrient profiles, lower levels of toxic metals, and—of course—no chemicals from pesticides[324]. All around, this means these are healthier for us in comparison to their non-organic alternatives.

Typically, mass-produced foods grown commercially are non-organically grown. To diminish waste and maximize profits, keeping pests away is vital and easily achievable with pesticides. The addition of artificial fertilizer boosts yield. With the added impact that large-scale agricultural land use has on our environment, better ways of growing our food are needed.

With this basis, let's look at some important matters about how we grow our food and how we could be growing our food in the future. We will explore the options of locally growing and internationally growing food: agroforestry, regenerative agriculture, tilless agriculture, and advancing innovation.

7.1. Local Food Supply Versus International Trade

There is a big argument for eating more locally grown foods than international ones. Benefits include greater nutritional content (farm-to-table) than some of the international food alternatives, heightened flavor, support for the local economy, promoting a safer food supply, and greater ease of tracing or finding out how the food was grown[325]. Earlier in the chapter we touched upon how internationally traded food makes up 19% of calories the average person consumes worldwide[326]:

> *"When the entire food supply chain was considered in this analysis the researchers found that global food miles equate to about 3.0 gigatonnes of carbon dioxide equivalent (GtCO2e) – higher than previously thought… The transport of fruit and vegetables contributes 36% of food miles emissions – around twice the amount of greenhouse gases (GHG) released during their production. Food miles only contributed 18% of the total freight miles, but the researchers found that the emissions from*

these made up 27% of total freight emissions, mostly from international trade (18%)." Directorate of Environment (2023)[327]

Eating locally may be challenging particularly when there is not enough diversity in your local ecosystem, seasonal limitations, or simply because of your personal dietary preferences. When you consider how some food simply cannot grow naturally in certain parts of the world, it compounds the difficulty of trying to eat locally, eat healthily, and still benefit our environment. Let's look at one example, the avocado, a food we do not find locally grown:

"Green Gold: The Avocado"

*"Avocado must be a magical fruit.
The name itself sounds like an invocation."
— Michael Bassey Johnson,
Song of a Nature Lover*[328]

Like others who enjoy the many health benefits of the creamy and nutrient-dense avocado, it's an enjoyable part of our dietary plan. We love it in our guacamole, as part of a balanced meal, on whole grain toast, as a dairy-free replacement for mayonnaise, or as part of a healthy dessert. The whole-food plant-based diet that my family eats is all the richer because of the inclusion of avocado in our meals.

As avocados are not locally grown here, we are left with the choice of eating internationally traded avocados or eating diets without them. Whilst exploring the impact of food systems on the environment, we came across some shocking facts: avocado contributes a whopping 128,478,412 food miles, averaging at 4,801 miles[329].

Not only are the GHGs emitted from transporting avocados concerning, but the rest of the environmental impact that comes from growing avocados is problematic. This includes the usage of fertilizer and pesticide, resources used for packaging, and the energy expended to process them[330]. This is without even discussing deforestation linked to avocado farming in countries like Mexico, further contributing to our problem of climate change due to the fewer forests to capture CO2 emissions[331].

In light of all these negative factors, we are faced with some uncomfortable choices: Do we stop eating avocado or ignore the facts? One of the premises that my family and I live by is "Once we know better, we can do better." Even though avocado still has a 50% lower carbon footprint than that of bacon, and sometimes even eggs[332], there is still room for us to do better.

Armed with this information, when possible, we can opt to find avocados that are sustainably grown and have lower food miles. Avocados farmed from Spain are a more environmentally friendly option for us than those from further afield[333]. We now know we have the option of freezing our avocado once perfectly ripe for up to three months[334] and ensuring we do not contribute to food waste which also contributes to CO2 emissions. Does it take more work to eat more consciously to care for our planet? Yes, definitely. But, wouldn't you say it's worth it?

- Ulrike

7.1. Regenerative Agriculture

Many indigenous communities around the world have practiced regenerative agriculture for centuries. But what is it? It is a method that addresses farming from a holistic perspective where man's needs for food are taken care of, and the environment is too by restoring soil and the health of the ecosystem, addressing inequality, and leaving our land as a worthy inheritance for generations to come[335]. Did you know that 80% of the world's biodiversity is in the world's remaining indigenous lands?[336]

Civilizations that practiced regenerative agriculture indigenously include the Lenca of Honduras, the ancient Maya people of South America, and some indigenous peoples of India and the Himalayas[337].

7.1.1. Agroforestry

Agroforestry falls under regenerative agriculture. It is the *"term for land-use systems and technologies where trees and other woody perennials are deliberately utilized in agricultural processes with other plants and or animals for their ecological and economic benefits within their ecosystem"*[338]. Essentially, with this form of agriculture, trees—especially perennial—and farm crops grow together on the same land.

With the world's booming population, the fact remains that enormous amounts of food still need to be grown to feed us all. Agroforestry could be the direction commercial and subsistence farmers alike can take more advantage of to create more biodiversity, a happier climate, and more harmony for our environments whilst still meeting our growing needs. The benefit of agroforestry over conventional farming where whole fields are dedicated to planting one crop is the buffering of crops from climate extremes[339], decreasing atmospheric carbon as these trees absorb them, reducing GHGs, and lowering energy usage on farms[340].

From an economic perspective, agroforestry holds even more benefits for humankind as it may offer the potential for multiple livelihoods, help regulate water and sediment flow, and carbon and nutrient integration into the soil contributing to better fertility of the soil, less soil erosion, and decreased need of pest control[341]. This may lead to higher crop productivity and naturally more profits.

7.1.2. No-tilless Agriculture

Like many regenerative agricultural practices, tilless agriculture is nothing new and was practiced as far back as 10,000 years ago[342]. Tilling is the process where land is prepared for growing crops[343]; tilless agriculture is therefore the type of agriculture that does not include the preparation of soil for planting or growing crops[344].

When soil is tilled, studies tell us that it impacts the distribution and transformation of organic carbon in the soil, aggregate stability (how firmly soil stays bound together even against other forces like rain), water holding capacity, soil temperature, and water exchange pathways between the soil surface and the atmosphere[345]. Tillage disrupts the soil microbial composition and contributes to changes in CO2 emissions from the soil[346]. Over the last 150 years about half of the world's topsoil has been eroded[347].

No-tillage agricultural practices offer a variety of benefits such as saving time, improving soil health, and saving on fuel costs. One study conducted in the US painted a very clear picture of the fuel savings farmers enjoy as a result of using no-till methods:

> "On average, farmers practising continuous conventional till use just over six gallons of diesel fuel per acre each year. Continuous no-till requires less than two gallons per acre. Across the country, that difference leads to nearly 282 million gallons of diesel fuel saved annually by farmers who practice continuous no-till instead of continuous conventional till.
> Farmers who manage at least one crop in their rotation without tilling – seasonal no-till – save an additional 306 million gallons of fuel annually."[348]

No-till agriculture does have its negative side such as the initial cost of changing agricultural methods, the learning curve for farmers, the possibility of chemical build-up gradually, and the formation of gullies over time[349].

7.2. More Innovation

We can live in harmony with nature and our Earth through the application of more innovation and technology in the way we grow our foods. Thankfully, scientists and technologists are hard at work trying to solve these complex and pressing problems. Two areas that hold promise that I would like to highlight, and that would lower the CO2 emissions of our food and food systems concern our mobility (which we will discuss in detail in Chapter 4) and innovative ways in which our food is grown—developments by Novo Nordisk Foundation in conjunction with the Bill and Melinda Foundation[350].

7.2.1. Green Mobility

When we make significant gains in clean energy as it relates to international mobility, the carbon footprint of our foods (such as our beloved avocado) will go down, helping us to enjoy it that much more than we can right now in light of the food miles the average avocado comes with. Just imagine all the produce you need getting to your doorstep on a zero-emissions mode of transport from the other side of the globe. We will see in Chapter 4 that progress in clean energy as fuel for transportation is being made which will hopefully lower CO2 emissions from our food systems.

7.2.2. CO2 as a Sustainable Input for Food Production

Scientists have started wondering: instead of CO2 being a problem, can it be part of a solution? Could CO2 emissions that end up damaging our climate be used to grow and produce food through fermentation processes (and not in soils or fields) that feed the world? Scientists have found a way to use CO2 as a raw material in the synthesis of protein for human consumption. Mid-2023, some exciting news from the Bill and Melinda Foundation, Novozymes A/S, and Topsoe A/S was released. They intended to address two pressing problems we are aware of—food insecurity and mitigating climate change—without involving agricultural land use, and whilst still ensuring foods grown are nutritious.[351] How do they intend to do this? Through the production of proteins through fermentation—not involving extensive land use:

> *"The basic idea is to provide a more sustainable way of producing proteins through fermentation – a way of producing food we have been using for millennia.*
>
> *By using biological and electrochemical processes, the consortium partners will process CO2 and turn it into acetate, which is vinegar – a well-known substance already present in the metabolism of the microorganisms used for fermentation. The acetate can then be used to produce proteins that can be used directly in food for humans.*
>
> *By creating alternatives to animal proteins, we can reduce the need for meat and dairy production, which puts a significant strain on our natural resources by using land for the animals and growing crops to feed them. In addition, using acetate derived from CO2 directly in the fermentation process will eliminate the need to use sugar, which is a big part of the fermentation process. This will free up substantial agricultural areas currently used for sugar production.*
>
> *Thus, converting CO2 into acetate and using it to produce proteins for food will enable us to decouple part of our food production from land use and make room for biodiversity. This will be a major contribution to a more sustainable society."*[352]

8. Holistic Wellbeing

Isn't it delightful when we feel strong, our mind is clear and nothing is perpetually nagging us in the recesses of our mind? Feeling well in our body, mind, and emotions is the state we would all like to maintain. It is very hard though to accomplish this level of well-being when one area is out of balance. In this section, let's take a moment to look at the linkage between diet and overall well-being, dopamine management, and some practical steps that we can take to reach a level of balance holistically.

8.1. Linkage Between Diet/Lifestyle and Well-being

When we are battling a disease, it's hard to enjoy good mental and emotional health. Research says that 60% of diseases can be avoided by eating healthy diets[353]. These chronic diseases include—as stated earlier—obesity, cardiovascular syndrome, some cancers, hypertension, stroke, diabetes (type 2), metabolic syndrome, and some neurological diseases (such as epilepsy)[354].

We've looked at how our appetites are impacting our overall health with increasing problems like obesity and overweight. We've also seen how global demand for food is contributing to roughly 30% of the global CO_2 emissions[355]. What if our issues aren't just about pesticides, harmful chemicals in our food packaging, and hormones exposed to our foods that make it harder for us to lose excess weight? What if part of the challenge leading to obesity and unfavorable health outcomes is an imbalance in the parts that make us whole: our minds, bodies, and souls?

8.2. Dopamine Management

You know that wonderful sensation you get when you eat your favorite chocolate bar, win a game, get a delightful message on social media or revel in rapturous applause after public speaking? That's you feeling the dopamine in your body. Dopamine is a neurotransmitter that sends messages between your nerve cells and plays a part in your body experiencing pleasure[356].

What has dopamine got to do with food, you may wonder. Well, dopamine is involved in experiencing food cravings[357]. Junk food, particularly ultra-processed food, is designed to keep you craving more because this food stimulates your reward system (dopamine neurotransmission system) and can lead to a clinically diagnosed food addiction[358]. One registered dietitian put it this way: *"From an evolutionary perspective, it makes sense that our bodies would crave the foods that have tremendous amounts of added refined carbohydrates and fats – it means more fat storage for survival with less effort." Carlos Fragoso*[359]

8.3. Taking Better Care of Ourselves

Dopamine is a hormone. If our dopamine levels are unbalanced, this is a problem that medical experts can assist us with balancing. A healthier future for both us and the planet will require us to take better care of ourselves. I am not sure who said it but I believe it's true: awareness is 50% of the solution. If we are to grow in consciousness and awareness, it may require some exercises that encourage the development of our inner selves, and gaining self-control and maybe even inner peace. These practices that may help us include meditation, self-hypnosis, mindfulness practices, and even autogenic training.

Meditation is a practice from ancient times that revolves around establishing and maintaining a calm mind and body during exercise[360]. Some studies suggest that practicing mindfulness lowers emotional and binge eating[361].

Self-hypnosis is the self-directed induction of a level of consciousness; a "state of deep relaxation and focused concentration"[362]. In this state, you can tap into your subconscious mind and receive and suggest a level of subtle guidance towards thoughts and thought patterns that may serve you better.

Autogenic training is another type of technique you can explore. It involves relaxation and focusing on increasing feelings of calmness and well-being[363]. You have probably heard of cortisol, also known as the stress hormone; our body secretes it when it surmises that you are under imminent threat[364]. Autogenic training is a viable option for those of us trying to manage unhealthy levels of stress, here's why:

> *"The relaxation technique uses autosuggestion to help a person perceive heaviness and warmth in a part of their body while releasing a slow breath. "Autosuggestion" refers to the use of words, statements, or cues to help guide one's own thoughts, feelings, or behaviors…The technique activates a part of the brain that may help promote and regulate self-healing mechanics in the body. In doing so, it may help the body self-heal from several different conditions, such as stress, trauma, or anxiety…"* Medical News Today[365]

A trained and trustworthy practitioner in any of these practices can help you get started. Remember, the size of our growing obesity and overweight problem may require us to make changes in what we eat and even the amount we eat.

8.4. The Placebo Effect

Even though I've backed these unconventional types of practices with scientific evidence, I understand if you may be skeptical about any of them as effective ways to break unhealthy eating habits and improve your health. I understand entirely. I will, however, draw your attention to the concept of the placebo effect. Though I consider these practices legitimately helpful, the placebo effect may make the practice of them impactful. The placebo effect, as you might know, is when someone's health improves after utilizing a placebo or fake treatment. It is a scientifically proven effect, triggered by the individual's belief in the effectiveness of the treatment that they're administered[366]. Our medical experts and scientists continue to discover new things about the workings of the body. Sometimes it takes time for these findings to be considered fact. Here's a story that can help me illustrate this further.

The Father of Hand Hygiene

> *"Seeing is believing, but sometimes the most real things in the world are the things we can't see."*
> — Chris Van Allsburg,
> The Polar Express[367]

At the time I was thinking about these practical steps you and I can take, such as self-hypnosis, mindfulness, and autogenic training, to find holistic well-being. I knew that readers might find it a bit "woo-woo" for them. Michael and I hesitated to put this in a book on climate change. But I wondered, what if these were some of the practices that would help some of us change our eating habits and decrease our carbon footprint? What if not including these negatively affected how much I could help you?

As I researched this chapter, I recalled a story I learned during my nursing and dental training about a German-Hungarian doctor who lived in the 1800s named Ignaz Semmelweis. This story is why this section on holistic well-being is included in this chapter.

Semmelweis was born in 1818, trained as a physician, and worked in the First Obstetrical Clinic at Vienna General Hospital, in Vienna[368]. In 1846, he began working as an obstetrician. At the time, 1-in-100 women died during childbirth at this hospital[369]. As an obstetrician, he dedicated himself to saving the lives of birth mothers and their children during a time when the leading cause of maternal death in Europe was due to puerperal fever, which affected postpartum mothers.

His curiosity about why the maternal death rate of postpartum mothers rose to 7.5% after the hospital directed obstetricians and medical students to also conduct autopsies led him to discover the existence of harmful germs[370]. His findings were ignored for years, and considered ridiculous.

It took more than two decades before germ theory became generally accepted. Sadly, Semmelweis also experienced a very tragic end. Thankfully, his findings did gain traction and many lives have been saved since.

The lesson here is: Even if things like self-hypnosis, autogenic training, mindfulness, and meditation seem only recently started, they have adequate backing from modern science. It's worth remembering that sometimes things work that we still do not have all the answers to.

Even if things like self-hypnosis, autogenic training, mindfulness, and meditation seem only recently to have been adequately supported by modern science, it's worth remembering that sometimes things work that we do not yet have all the answers to.

As a trained hypnotherapist and nurse, I appreciate the words of experts in these practices. One such noteworthy scholar, a professor from Stanford, Dr. David Spiegel, (medical doctor and associate chair of psychiatry and behavioral sciences) champions the relevance of hypnosis. Dr. Spiegel states:

> *"Hypnosis is the oldest Western form of psychotherapy, but it's been tarred with the brush of dangling watches and*

purple capes. In fact, it's a very powerful means of changing the way we use our minds to control perception and our bodies… When you're really engaged in something, you don't really think about doing it—you just do it." Williams, 2016[371]

Numerous studies reveal how specific parts of the human brain are altered through the use of hypnosis[372]*. Even self-hypnosis has been proven useful for realizing goals such as weight loss*[373]*.*

Meditation is another popular practice, including among many famous and successful people such as the founder of The Huffington Post, Arianna Huffington, the billionaire and philanthropist Oprah Winfrey, and Candy Crowley, journalist and CNN presenter[374]*.*

Autogenic training, along with these other practices, has been cited in various studies and journals for its potential to make a positive impact on the lives of those who practice it. This specialized relaxation technique uses systematic exercises and has been found to relieve chronic pain in some patients and improve well-being[375]*.*

Mindfulness has also been shown to offer numerous benefits such as relieving depression, stress, and anxiety, all of which can impact one's dietary choices and overall health[376]*.*

If all this is new to you, it is worth noting that all these practices help you enter an altered state. The main difference with all these practices is what you do with an altered state. In meditation, you just observe. In self-hypnosis you just give suggestions. In autogenic training, you have a feedback mechanism that guides you. In mindfulness, you are very mindful of the now. It's best to find something that fits you.

These practices are scientifically proven and will help you be happier, have better well-being, and lead you to make a more positive environmental impact on the planet. All factors being equal, it's okay to use those practices that work, even if we don't know as individuals how they work when we know that they are doing us and the world some good. Just allow yourself to be open to changing your life for your benefit and the benefit of the planet.

- Ulrike

9. Recommendations

With such a vast topic, there may be no end to the recommendations we can make; hence, my list here is not exhaustive. Rather, consider it a starting point which you can evaluate for what's most workable for you. Some of these recommendations we've already alluded to such as the first; eating more locally grown foods.

9.1. Eat Predominately Locally Grown Food

Though this may not always be possible, where it is, it would benefit our climate more if we defaulted to a home-grown or locally-grown diet of organic foods. The nutrition content, the lower food miles, and the fact that we have the opportunity to know exactly what goes into the production of our food make for healthier and likely more sustainable daily diets in the long term.

9.2. Support Local Farmers

Organic farmers need our financial support as consumers of their produce, marketers, and even investors. If you have the resources to invest in your local food industry or advocate for local players, consider taking up the role. When faced with behemoth conglomerate enterprises as competitors, often the local farm does not stand a chance against the incredibly well-financed, effectively marketed, and exceptionally resourced Big Food industry. Eating local foods ensures we keep these local businesses in business, cutting those food miles off the carbon footprint of our diets and subsequently lowering our carbon footprints.

9.3. Reduce Food Waste: Eat "Funny" Fruit (and Vegetables)

This is one of the recommendations of the UN Sustainable Goals (Global Goals)[377]. Focus on the quality of the food and not the cosmetics of it. A misshapen apple is just as nutritious as a perfectly symmetrical one. Over-ripe bananas are excellent for desserts and whole-grain banana bread. Some perfectly good quality fruits and vegetables end up binned purely because of cosmetic reasons[378]. If we purchase and consume these fruits and vegetables, we lower food waste and the associated $CO2$ emissions that decomposing food emits.

In addition, freezing or preserving foods we know might spoil can help us further avoid food waste. Other than freezing our foods, other methods of preservation used for centuries around the world may be as beneficial or even more so such as pickling (which enriches foods with probiotics through the fermentation process)[379]. Plant-based foods that can be pickled include fermented cucumber, green olives, cabbage, garlic, onions, carrots, capers, and many other vegetables[380]. Asparagus, parsnips, radishes, peapods, small florets of cauliflower, and broccoli are all additional options[381]. You can also pickle fruits and all types of meats[382].

9.4. Eat Less Meat

We must move away from our preference for animal-based protein. It is killing us! We may never reach a consensus when it comes to our daily diets but one approach that could help us is by beginning to eat a plant-based diet once a week. This is another recommendation of the UN Sustainable Development Goals[383]. For us to see the long-term benefits, a shift towards a vegan or whole-food plant-based diet just once a week is not enough to support our planet against the growing impact of climate change. More of us must make the shift towards climate-friendly daily diets. Furthermore, we as individuals will reap the benefits: one major study showed that the replacement of 3% energy from animal-based protein with plant-based protein was inversely associated with overall mortality and heart-related diseases that cause death amounting to an 11% reduction of risk in men and a 12% reduction of risk in women[384]. According to the study, even just removing eggs from their daily diet caused a decrease of 24% in overall mortality in men and a 21% decrease in overall mortality in women.

One main criticism of the vegan or wholefood plant-based diets is inadequate amounts of complete protein, preventing for the body's optimum function. This is a misconception as there are a variety of foods you can prepare and consume that will give your body all the proteins it needs. To get complete proteins in your diet, sometimes this means putting together combinations of foods. Examples include: hummus and pita, buckwheat, soy (and soy-based products like tofu, tempeh, and soybeans), pumpkin seeds, peanut butter on toast, hemp, beans and rice, Ezekiel bread (made from sprouted grains), chia seeds, quinoa, and spirulina[385]. Apart from the complete proteins you get from these foods, you can also find significant quantities of Vitamin B12 in several plant-based foods such as nutritional yeast, cereal, mushrooms, and some types of algae[386]. I hope you see that switching to a plant-based diet may mean being more intentional and selective about what you eat but does not translate into deficiencies. Rather, long-term change translates to even better health for you and the planet.

9.5. Lobby for Humane Practices in Farming and Less Food Waste

Let's get vocal and demand humane treatment of animals farmed for food. If you continue to eat animal products, as much as I would love for you to switch to a plant-based diet, I think at the very least you and I can lobby for their reasonable and fair care. Cruel practices like debeaking, dehorning, molting, overcrowding, and starvation are practices that should be unacceptable in our modern era. We must push for authorities to eliminate them in favor of more humane practices.

We know that enormous amounts of food go to waste which could otherwise be provided to those suffering from food insecurity, leading to additional CO_2 emissions as those foods rot and go to waste. Making our voices heard by authorities and our favorite food enterprises about our concerns about food loss and food waste can help them make more effective steps and decrease their impact on this global problem.

9.6. Investing in Innovation

We must put more of our resources through credible institutions and initiatives towards solving our food-related problems, like land use for food production, food security, food waste, and decreasing the CO_2 of food. We need to invest in innovation just as the UPF industry does. Increased research and development on organic plant-based food production processes can make a difference. The magnitude of the economic impact our health has on us as a society and the opportunity it presents from a capitalist perspective is real. If you have the resources, investing in innovations to solve some of the problems we've outlined in this chapter could ultimately add to your wealth and solve some of the world's most pressing problems at the same time.

9.7. Educate Others

We can no longer allow ourselves to be ignorant in the age of information. Be curious about what goes into your food and educate others about the impact of the food they choose to eat. Find out how to reduce food waste and share with your social circle how they can keep their food edible and reduce food waste. In addition, the more people who advocate for the humane treatment of animals or know how much healthier organic food is for them over ultra-processed foods, the more likely they may make the healthier choices given the chance and resources. By sharing with others what you've learned in this book and elsewhere, you contribute to the solution.

9.8. Live a Healthy Lifestyle

Focus on creating your healthiest self. It will breed a healthier environment around you. There are many practices such as exercise, yoga, breathwork, and meditation that you can research and find out if they fit you. The bottom line is to find ways to curb the appetites that we have that contribute negatively to our health and climate change. For some of us, this may look like going to therapy to treat a food addiction. To another, this might be giving up ultra-processed foods or eating more plant-based foods. Over time, the positive changes you make will show. However, this does not happen overnight, changes like this need to be lifestyle changes and not temporary fixes for us to reap the maximum benefits.

10. Conclusion

Our food and the systems that help get it to us is a complex subject. Hopefully, at this point, you have come to see how both our food and our current systems impact climate change and have a few ideas as to how we can make a difference now wherever we are. Even as I wrote this chapter, I could easily see how some of the changes we would need to make would cost me and my family. I imagine you saw the same for yourself. As I conclude this topic, the one question that I asked earlier is worth repeating: Is the sacrifice worth it for the harmony we stand to gain?

In this chapter, we looked at how our food and food systems impact climate change, what is being done, and what we can do to lessen our negative environmental impact. In the next chapter, we will focus on our current transportation or mobility solutions and what an environmentally friendly future would demand from us to change.

Chapter 4
MOBILITY OF THE FUTURE

"Climate change knows no borders. It will not stop before the Pacific islands and the whole of the international community here has to shoulder a responsibility to bring about a sustainable development"
Angela Merkel[387]

In the previous chapter, we looked at food and food systems, and the changes we need to make to ensure we do better, going forward. In this chapter, we focus on the future of our mobility. Before we explore the concept of mobility in the future, we first need to agree on what we mean by the term "mobility." Then we will look at the contribution of CO_2 from transportation. We will take a moment to delve into that as well. Mobility in the city, autonomous cars, bikes and e-scooters, emissions from powertrains, and more innovations are topics we will take time to understand before looking at our key takeaways.

Think about the world in 15 or 20 years. How do you envision you will get around? Do you see yourself driving your car to work or being driven by a robotaxi as you hurriedly finish off your report due that morning or as you polish off your breakfast in the back of a car? Perhaps you have a more vivid imagination and can see yourself not being driven but traveling in a flying taxi, taking just a fraction of the time it would normally take you to commute. Outlandish? Perhaps. But did you know that a flying car, still in development, completed its first-ever intercity flight back on the 28th of June 2021 in Slovakia?[388]

So, let's begin by understanding the term "mobility".

1. What Is "Mobility"?

"Mobility" is "The potential for movement and the ability to get from one place to another using one or more modes of transport to meet daily needs"[389]. This includes the movement that you and I engage in to commute for work and school, to move animals, transfer physical items, and to conduct every aspect of living across locations. It also encompasses commercial purposes such as the transportation of goods and services or service providers. Maybe you travel by tram, train, car, or plane routinely. Perhaps biking or walking is your preference. Whatever the case, we all need to brace ourselves and prepare for mobility in the future.

My Paris Miracle

"You are not stuck in traffic. You are traffic."
- an ad by TomTom N.V[390]

Paris is a beautiful city and I've been fortunate to travel there countless times, many times for business. Like many bustling cities, it's plagued by many of the issues that high population density attracts, one of those problems being traffic. I remember one occasion, one evening when I had an international flight to catch out of Paris. I was at the office and knew that the two-hour commute in heavy traffic would lead to me missing my plane. As a stereotypical German, punctuality is etched in my DNA, and missing my plane was something I was not prepared to allow. Then it dawned on me: I could hail a ride. A motorbike!

I hopped on the back of it as the motorcyclist's passenger. Fortunately, I travel light for most business trips. Clad in my pressed business suit, gripping my travel bag and trying to hold on to the bike for balance—and dear life—I braced myself. We zoomed through the lanes of still traffic at speeds that made me hold my breath. I guess that was my doing since I had instructed the rider on how I needed to catch my plane and time was of the essence.

When I arrived at the airport, shaken but appreciative with enough time to spare, This ride that saved me a missed flight, 1.5 hours stuck in traffic, and a lot of stress had cost me 150 Euro.

I was grateful for the option of mobility that this rider and his two fast wheels had given me. I was also grateful that I was open-minded enough not to dismiss the idea as undignified or unrealistic. When I was up in the plane flying home, and the memory crossed my mind, I was grateful for the multiple types of mobility modern life offers in light of all our needs.

- Michael

2. Classification of New Types of Vehicles

Cable[391], space[392], roads, water, air, rail, intermodal, and pipelines are all modes of transportation for mobility[393]. Our focus is on the mobility of the future, and as such, we will use the classification presented by the International Transport Forum (ITF), an OECD intergovernmental institution tasked with offering insights and knowledge on transport policy. Their classifications for new mobility vehicles in their report titled *"Measuring New Mobility: Definitions, Indicators, Data Collection"* are micromobility, powered light mobility, and car and van-like vehicles[394].

2.1. Micromobility

The first classification for new vehicles presented by ITF, micromobility, includes manually-propelled bicycles, electric bicycles, scooters, and other rideable vehicles. They may have two or more wheels. In one study conducted across eight cities, looking at 3,800 commuters, the research showed that commuters switching from cars to bikes cut emissions by 67%![395] Imagine the gains we could enjoy if around the world we made the switch once a week or more? We will take a deeper look into micromobility vehicles further into this chapter.

2.2. Powered Light Mobility

Powered light mobility, L-category vehicles or powered light vehicles (PLV) encompass two-, three- and four-wheeled vehicles for either passenger or cargo use, ranging from mopeds to quadricycles[396]. This classification of new vehicles by the ITF includes throttled e-bikes, mopeds, e-mopeds, motorbikes, e-motorbikes, rickshaws, and e-rickshaws. We will take a further look at powered light mobility later in the chapter.

2.3. Car and Van-like Vehicles

The third and last main classification for new mobility vehicles by ITF includes both cars and van-like vehicles. Cars with internal combustion engines (ICE) and private van-like transportation are one sub-classification of this third group. Cars that have internal combustion engines use fuel and an oxidizer to create a reaction called combustion[397]. This process releases the energy that mobilizes the vehicle. Among the fuels that can be used in this combustion process are biogas, hydrogen, and various types of fossil fuels[398].

The hybrid car or van is a vehicle that *is "powered by an internal combustion engine and an electric motor. These greener vehicles allow drivers to experience their more improved fuel economy than conventional cars."*[399] A battery-electric car or van has an electric motor only[400]. An autonomous car or van (automated car or van/self-driving or driverless car) *"employs driver assistance technologies to remove the need for a human operator"*[401]. Lastly, we have the minibus or microbus or jitney.

This is a *"motor vehicle designed to carry a moderate number of passengers, typically between 12 and 30 people, used by private and public transport services"*[402].

3. CO2 of Transportation

At present, 1.3 billion vehicles globally are utilized, with the majority being owned privately[403]. From 2021 to 2022 GHGs increased by 3%, driven by an increase in air travel after the pandemic[404]. This gradual increase is due to the resumption of virtually normal air transportation post-COVID-19 pandemic era. As a seasoned industry player in the transportation sector, coupled with my interest in climate change matters, I've discovered that many people overstate the impact that the transport sector has on climate change. If you can recall from Chapter 1, we saw that the industry and power sectors all contributed significantly more CO2 than the transport sector. Perhaps the transport sector is perceived to be far more negatively impactful because we see smoke blowing out of trucks and cars on our roads with a frequency that gets our attention and drives the perception. Not everyone lives next to industrial factories or sees billowing smoke rising from their industrial chimneys. Or if we do, perhaps we've become so used to the sight that, like a chameleon, these mammoth structures have stealthily blended into our background.

4. Mobility in the City

Getting around in cities is another facet of mobility of the future. Luckily, we have some idea of what we can all look forward to in that regard. With new types of transport emerging like autonomous vehicles, and the evolution and innovation that technology and demand are driving, there are already new services that can help us understand what we'll need to prepare for as we go about our days in locations around the world. As more of us become aware of all the options available and more transport businesses spring up, we can expect to see more services springing up to meet the demand that we drive. A growing business model dubbed "mobility as a service" (MAAS) is part of the equation and innovation in this sector[405]. Here's a diagram from the International Transport Forum categorizing new mobility services that can help us see how comprehensive our options already are in some cities around the world:

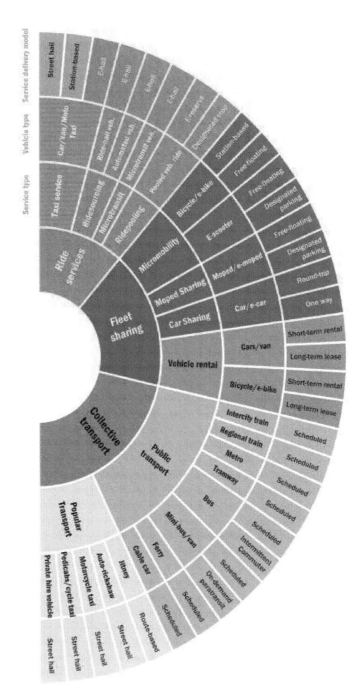

Credit Source: International Transport Forum. "Measuring New Mobility: Definitions, Indicators, Data Collection"[406]

As you can see in the diagram there are three main categories of new mobility services: ride sharing services, fleet sharing, and collective transport. Let's take a closer look at each.

4.1. Ride-Sharing

Needing to get somewhere doesn't necessarily mean you need to own your own vehicle. Your living conditions, lifestyle, and even location could make owning a vehicle impractical and unnecessarily expensive. With new mobility services like ride-sharing, you could contact a ride-sharing service of any of the four types: taxi, ride-sourcing, microtransit, and ridepooling.

Taxi services are something most of us are familiar with. They are commercial services that transport a passenger or passengers by road from one location to a taxi stop or requested destination. Some of the world's most popular are Bangkok's tuk-tuks, Kolkata's Ambassadors, Mexico's Vochos, Shanghai's VW Jettas, Moscow's Porsches, Hong Kong's Toyota Crown Comforts, Nürburgring's BMWs, Havana's jalopies, New York's yellow cabs and London's hackney carriages[407].

Have you ever used a third-party service through a website or an app that allowed you to request or hail a ride that you paid for? If so, then you've used a ride-sourcing service. This is another type of ride-sharing service. Uber, Ola, and Bolt are some of the world's leading ride-sourcing companies[408].

Another type of ride-sharing service is "microtransit". Think "Uber" or "Lyft," but instead of the vehicle being a sedan, it's a bus. Microtransit involves agencies providing riders with an on-demand bus service that is inherently more flexible and does not have fixed routes and fixed departure and arrival times[409]. Companies offering this type of service often integrate their services with public transit to give users a seamless transportation experience if they choose to connect to another transport service. Microtransit services can be geared towards private users or public users. Companies like TransLoc OnDemand, Via or the transportation services that some companies provide their employees are good examples of this[410]. Along these lines, Google has a shuttle service for their employees.

Lastly, among ride-sharing options, there is a type called "ridepooling," which is a commercial service that involves bundling several people with their travel requests into one vehicle for their joint transportation and drop-offs[411]. Lyft, Uber, and MOIA in select German cities are some ridepooling service providers[412].

4.2. Fleet

Fleet sharing is the personal usage of permitted licensed vehicles by users at times that they book the service[413]. Fleet service types include micromobility, car sharing, moped sharing, and vehicle rental[414].

Micromobility fleet services typically offer bicycle/e-bike or e-scooters, which can be collected at stations or—if they permit—anywhere on the street, in the case of dockless or free-floating systems[415]. Mobike in Beijing and Jump e-bikes are examples of providers of this service[416].

Moped sharing is another type of fleet-sharing service that allows users to rent mopeds or e-mopeds and collect or drop them off at designated parking spots or if permitted on an undocked/ free-floating basis[417].

For a fee, car- or e-car-sharing fleet services let users rent a car without restricting their usage to pick-up locations or designated office hours[418]. Some companies offer this service in a club model. Examples of companies offering car-sharing fleet services include Avis, Enterprise Rent-A-Car, and Europcar[419].

Another service under fleet sharing as categorized by the International Transport Forum is vehicle rental. These services provide the rental of bicycles, e-bikes, and/or cars and vans for users at a fee[420], typically on a short-term basis or long-term lease option. Some of the world's leading vehicle rental companies are Avis Budget Group Inc., ALD SA, Enterprise Holdings Inc., Hertz Global Holdings Inc., and Sixt E[421].

4.3. Collective Transport

There are two options: public transport and popular transport.

4.3.1. Popular Transport

Types of popular transport include private hire vehicles, pedicabs/cycle taxis, motorcycle taxis, auto-rickshaws, and jitneys[422]. All, except for jitneys, are services you can hail randomly on the street[423]. Jitneys (a type of bus or other type of vehicle) typically follow a pre-destined route[424].

4.3.2. Public Transport

In many places around the world, public transport offers a variety of options. This is mainly the case in vibrant cities such as Paris, New York, Cape Town, Rio de Janeiro, London, and New Delhi. Vehicle types for this type of transport include cable cars, ferries, mini-buses or vans, buses, tramways and underground rail, as well as regional and inter-city trains[425]. For most of these vehicle types, service delivery is typically on a scheduled basis. Some minibuses/van public transport services offer on-demand paratransit instead of just scheduled options. Additionally, some public bus services allow for intermittent commuting[426].

4.4. Urban Development

Mobility of the future relies heavily on good urban development and design. Urban development is "the construction on land of improvements for residential, institutional, commercial, industrial, transportation, public flood control, and recreational and similar uses"[427]. Mobility of the future requires exceptional attention to urban development and design that helps us reap climate-friendly benefits. In other cases, some cities have still managed to leverage existing design by incorporating innovations that still make the existing infrastructure work for—and not against—a climate-friendly agenda.

Of course, most countries that have been able to implement and accomplish impressive results all have significant financial resources to make this a reality. These countries include the United Kingdom, Norway, Germany, France and the United States[428]. It's probably realistic to assume that countries and cities without such budgets would need unique strategies to mobilize financial resources to achieve the same. Among the world's 10 most congested cities, commuters lost an average of 79 hours stuck in rush hour traffic in 2022[429]. Of these 10 cities, eight were in Asia and Latin America, with Bogota city in Colombia ranking as the most congested by TomTom Traffic Index data[430].

4.5. Expected Trends

As we speak, the world is in the throes of change, and more change is expected. Let's look at a few of the trends we can expect or that have already started unfolding.

4.5.1. Decrease in Private Vehicle Usage

One study found that many cars in cities around the world are left parked for close to 95% of their private ownership[431]. Mckinsey estimates that by the year 2035 private usage of cars will drop by 15% globally[432]. Their Future of Mobility April 2023 report presents expectations of newer types of mobility and services (public transportation, e-scooters, e-bikes, and smaller more efficient cars) gaining global market share from their current 1% to 8%, enabling this change[433]. The projection is that these changes will occur at a slow pace as meanwhile most of us use the transportation options we've reverted to all along.

Globally, rural areas will lag in the adoption of these technologies, whilst cities known for their progressive stances such as Shanghai, Munich, and Los Angeles lead the way for us all. This part is a projection that some of us can debate: though China leads in adoption and even investment of new mobility, the US market as a whole has lagged behind projections set for electric vehicle adoption in 2023, raising significant concerns about how this impacts the globe as a whole[434]. This is a legitimate concern, as though electric vehicles are only part of the solution, the usage of these vehicles in new mobility services is the combination that can help the US and the world at large enjoy the larger environmental gains.

The Mckinsey report of April 2023 also estimates a continued rise of car sales worldwide with a peak at the end of the decade, followed by a decline to 84 million units in 2035 led by Europe and the US markets[435].

4.5.2. RoboTaxis and RoboShuttles

Have you ever heard of robotaxis and roboshuttles? Well, these machine-driven vehicles are projected to meet the mobility service demands of the future. We will take a more concentrated look at autonomous vehicles in the next section of this chapter.

Technology and the businesses that utilize it are only viable if there is a market and demand for these new mobility services and solutions. If demand for cleaner and more sustainable mobility solutions continues to grow, it may become the biggest driver behind all this change.

4.5.3. Increased Ride-sharing

We looked at ride-sharing as a category of new mobility services earlier in this chapter. This sector is likely to grow as an industry as more of us turn to ride-sharing for our transportation needs and investors and entrepreneurs see the potential of this booming sector and take their chances as being the next Uber or Lyft in their areas, especially if they are in countries or regions that are significantly under-served. As you might know, companies like Uber and Lyft have huge market shares. In countries where there is no Uber, it's reasonable to assume the potential viability of offering a similar service in its stead. In the past, Uber and Lyft have faced suspensions, been banned, or will be banned in some countries respectively for failing to meet regulations. Denmark, Italy, Bulgaria, Hungary, Finland, France, Spain and the Netherlands are among some of these countries where Uber has seemingly faced challenges meeting regulations[436]. Lyft, on the other hand, only operates in Canada and the US[437].

4.6. Regulations

Another very important facet of the change we should all anticipate in the coming years is that of changing regulations in cities worldwide. In response to the GHG emissions problem and challenges presented by private-vehicle congestion on road networks, many governments (150 cities worldwide) have instituted regulations intended to reduce the number of vehicles on their roads[438]. Some governments are also providing financial incentives for private-vehicle owners if they decide to use cleaner modes of transport. Here are examples from the Mckinsey publication[439] of some of the regulations being implemented around the world towards mobility of the future:

- Norway: the government established a Climate Change Act targeting to reduce their national emissions by more than half by 2030, and to become a low-emissions nation by 2050. The nation has also made some changes in terms of urban development by providing more bike lanes and pedestrian roads.

- China is constructing the world's largest urban cycle lane network. Their plan is for it to have a maximum length of 1,920 km by 2025 and 17,000 km by 2040. The nation is also expanding its subway system by 1,625 km, with a goal of rail transport amounting to 27% of the nation's public transit.
- United States: The US instituted the Bipartisan Infrastructure Law in 2022 which stipulates the provision of $1.44 billion each year towards investment and improvements related to pedestrian and bicycle infrastructure. This is a significant increase from their prior provision in 2018 of $850 million.

If these trends in regulations spread to more countries and cities, we can safely expect to find it more practical for ourselves to use e-vehicles or shared public transportation to go about our day.

5. Autonomous Vehicle (AV)

We've briefly touched on the concept of autonomous cars (robotaxis and roboshuttles). In this section, let's take a more concerted look at this new form of mobility. This is a type of vehicle that uses several inbuilt systems such as radar, cameras, GPS, and monitoring systems to transport commuters without dependence on a human driver[440].

5.1. Levels of Automation

AVs are not all the same. The differentiation is based on the levels of technology that AVs will have to progress through. The Society of Automotive Engineers (SAE) designates six levels that differentiate AVs[441]: Level 0 (no driving automation); Level 1 (driver assistance); Level 2 (partial driving automation); Level 3 (conditional driving automation); Level 4 (high driving automation); and Level 5 (full driving automation). The following is a roundup adapted from Synopsys[442]:

In terms of automation, the cars that most people drive are at level 0 as of 2023. This ranking indicates that a vehicle has no automation. For that reason, the lowest level of automation is Level 1: partial driving automation.

ADAS or advanced assistance driver systems: Level 2 (partial driving automation) are vehicles that can steer and regulate speed without the aid of a human driver. Cars at this level of automation still allow for a human driver and give them the option of taking control of the car at any point during the drive.

Conditional driving automaton: Level 3 SAE has environmental detection technology built into the vehicle. These types of cars can speed up or slow down to navigate through traffic. Other than this, these Level 3 AVs still need a human driver to override the driving experience, requiring the drivers to pay attention at all times just in case they need to take control in a fraction of a second.

Level 4 SAE: high driving automation has a significant difference from Level 3 because if an emergency arises during a drive, the car does not need human intervention. The vehicle can—in most cases—take the necessary actions. As of now, such cars are limited to a maximum speed of 30 mph or 48.28 km/hr. This is by legislation due to the infancy of this technology and concerns of public safety.

Level 5: full driving automation requires zero human intervention; most car manufacturers are vying for this level of technology[443]. As such, these vehicles will not include steering wheels or pedals. Once fully integrated into our societies, they will not even be limited to the 30mph speed limit placed on Level 3 vehicles.

5.2. Where Are We Now?

As stated earlier in the book, the world is many years away from reaching Level 5 SAE automation and is not likely to achieve it by 2030. You can identify Level 1 cars by whether or not they at least include technology for adaptive cruise control and lane-keeping assistance[444]. Several car brands integrate adaptive cruise control, lane-centering, stop-and-go capabilities, and route information to improve the driver's experience of the journey; these are all examples of Level 2 SAE AVs[445].

A good example of Level 3 SAE is Mercedes-Benz's vehicles that have their proprietary Drive Pilot system[446]. The company Waymo offering their automated driving capabilities in their taxis available in select cities in the United States can be considered Level 4 SAE AVs[447]. If you are wondering what level of automation Tesla cars have, as of 2016, this company's cars were reported to all have at least Level 2 automation[448].

The future is promising. Like anything, AVs will offer some advantages and disadvantages once they are mainstream in society. Let's take a look at some of these:

5.3. Advantages of AVs

- **Safety:** Driver behavior and human error causes 93% of all road crashes[449]. AVs won't have humans as a variable. This translates to greater safety on our roads and lives saved to the tune of almost 1.3 million[450].
- We never need to worry about our Level 5 AVs one day getting drunk, texting as they drive, or getting distracted as they transport us wherever we need to go. Consider the moments you or someone you love has driven home drunk. Some of us have lost loved ones because of this. If Level 5

AVs existed long ago, maybe our loved ones would still be with us.

- **Provide mobility support for people who need support**[451]**:** The elderly, and others with mobility constraints, need to benefit greatly from the usage of AVs in the future.

- **Free Up Time:** Imagine using your time for driving to have undistracted time catching up with your loved ones over the phone on the way to work, studying, or simply relaxing. If you had to drive, the split concentration would not have been safe.

- **Free Up Funds for Other National Needs:** According to the World Health Organization (WHO), road traffic injuries cost most countries approximately 3% of their gross domestic product (GDP)[452]. GDP is the "total monetary value, or market value, of finished goods and services produced within a country during a period, typically one year or quarter."[453]

- **Bridge the gap of shortage of truck drivers:** Do you know that there's been an ongoing shortage of truck drivers globally? A report in 2021 showed a global shortage of 2.6 million truck drivers across all regions[454].

5.4. Disadvantages of AVs

- **Disrupt the Transport Jobs Market:** When AVs take over the truck driving and commercial driving job markets in the future, we can expect jobs to be lost to technology. The encouraging understanding we all need is that even if machines take some of the jobs we've traditionally done ourselves, this new technology will open up new job opportunities. Retraining, staying informed, and preparing to align with the future is the best way to project one's job or future livelihood.

- **Potential of Failure:** Just as we have human error, the potential for AV to fail is a possibility. After all, humans made it. This is one of the leading arguments by critics of AV technology. This is a legitimate concern. We all need to keep in mind that if these AV companies get it right, the solutions they offer may still decrease the number of lives lost through human error each year. And I would argue that each life saved is worth the switch to AV technology.
- **Cost:** Developing new cars on cutting-edge technology is expensive. A Quartz article once revealed just how expensive developing AVs is:

  "Amid a race to market, the promoters who say a self-driving future is just around the corner rarely disclose one important thing—the price of outfitting an autonomous car. That extra cost, according to one of the few experts prepared to discuss the subject openly: is about $250,000 per vehicle.

  When you ask carmakers and industry researchers about the cost of self-driving equipment, they almost always say around $8,000–$10,000.[455]"

- **Adoption Rate with Drive Efficacy of AV Integration:** Coupled with the uncertainty of when Level 5 AVs will roll out for us to utilize and the skepticism and concerns from the potential markets, the initial projection of AV adoption is extremely conservative. McKinsey projects several scenarios, with one of them projecting a 4% Level 5 AV adoption rate in 2030, rising to 17% in 2035[456].

As an expert in the automotive industry, I'm quite eager for the benefits that AVs will give societies around the world. Many of us in the sector hoped to see this technology far more advanced by now. However, many key players in the space have since adjusted their projections to a more conservative stance with the acceptance that we are not likely to have AVs available as a normal travel option anytime soon, certainly not by 2030[457]. Are you wondering why that is? A comment purportedly by one leader in the space, Axel Schmidt from Accenture sums up the complexities of the technology well:

> "By now many expected that we'd see robotaxis driving around, but the problems are very complex. Driving on the highway when all the cars are going in the same direction is much easier than handling inner-city traffic. And much of the early research was done in California where there's not much rain, fog or snow. In Europe's all-weather conditions the problems were all a bit underestimated with things like snow blocking road markings." Axel Schmidt[458].

6. Micromobility and Powered Light Mobility

Earlier in this chapter we looked at the classifications of vehicles. We saw that many options exist for us to take advantage of, from one wheel to four: manual, electric hybrid, and autonomous. We can all agree: not every mode of mobility is practical for daily use—for example the skateboard or unicycle—or feasible (e.g. Level 5 AVs) for the average commuter. Some types of vehicles are just for sport or fun. So in this section we want to focus on the more practical smaller vehicles classified under micromobility and powered light mobility that present us with realistic, environmentally friendly, and healthier options for our daily commuting needs. With that said, let's jump into understanding all our micromobility options further.

6.1. Micromobility Vehicles

Imagine not having to worry about traffic because the bike lane is not congested. You manage to whizz to work without the frustrations and inconveniences that come with noisier, slower, and emotionally destabilizing inter-city travel that leaves you rattled as you start your work day. You get to work energized because your 15-minute bike ride had your blood pumping. You don't need to hit the gym if all you're trying to do is maintain your weight because your daily commuting is helping you burn that extra fuel and some of the cortisol your body naturally produces every morning. Your waistline is noticeably smaller and people notice the difference.

In cities where many people use micromobility, they also enjoy less air pollution and less noise pollution, further improving society's quality of life. This small change in lifestyle choices at an individual and collective level can have such a knock-on effect of benefits that you and I can enjoy when weather and other factors affecting micromobility are favorable.

For that reason, this classification of vehicles is of extreme importance in the discussion of making a positive environmental difference. With all this said, we need to know what our micromobility options are.

6.2. Micromobility Vehicle Types

In this classification by the ITF, we have manually propelled bicycles, electric bicycles, scooters, and other rideables. Manually propelled bicycles include pedal-propelled bikes, cargo bikes, and pedicabs. A cargo bike is "a *bicycle designed for transporting loads, with a large container attached to it that sometimes has its own set of wheels*"[459]. A pedicab is *"a bicycle with three wheels, with a covered seat at the back for passengers, used as a form of taxi"*[460]. An electric bicycle or e-bike is *"a bicycle equipped with an electric motor that may be activated in order to assist with or replace pedaling"*[461].

Types of e-bikes available in the market are the pedelec, speed-pedelec, and e-pedicab. The pedelec is *"a bicycle with an electric motor that helps the rider turn the pedals"*[462]. The speed-pedelec, on the other hand, is *"a type of pedelec (= a bicycle with an electric motor that helps the* rider turn the pedals), on which you can travel at a higher speed than you can with a basic pedelec An e-pedicab is the electronic version of a pedicab.

Another type of micromobility vehicle is the scooter. The term "scooter" emanates from the design of the vehicle which has a flat frame including a foot platform[463]. Standing scooters, e-scooters, and mobility scooters are all types of scooters. A standing scooter is *"a vehicle ridden while standing that consists of a narrow footboard mounted between or atop two wheels tandem that has an upright steering handle attached to the front wheel, and that is moved by pushing with one foot"*[464]. Depending on the model, a standing scooter may have two-to-four wheels. An e-scooter or electric scooter is a "scooter propelled by an electric motor"[465]. When people have physical challenges or limitations moving shorter distances than more robust vehicles would be more suitable, mobility scooters are useful. A mobility scooter is *"a vehicle ridden while seated that usually has three or four wheels, is typically propelled by an electric motor, and is used by those with impaired mobility"*[466].

"Other Rideables," another classification under Micromobility are typically miniature vehicles engineered to be used by one person at a time[467]. These include the skateboard, electric skateboard, hoverboard, one-wheel, and electric unicycle. You are likely familiar with the skateboard. If not, it is *"a short board mounted on small wheels that are used for coasting and for performing athletic stunts"*[468]. There are now electronically powered skateboards called electric skateboards. A one-wheel vehicle is a *"self-balancing electric personal transporter on which the user stands and places feet perpendicular to the direction of travel on front and back platforms"*[469]. An e-unicycle or electric unicycle (yuki/uni) is a *"self-balancing electric personal transporter with a single wheel. The rider controls the speed by leaning forwards or backwards and steers by twisting the unit using their feet. Some dual-wheel models exist, but the principle remains that of a single axle device used with feet in the direction of travel"*[470]. Electric skates or e-skates/powered skates) are a *"pair of skates with electric batteries and motors, controlled by the user leaning forwards or backward or using a remote control"*[471].

Science fiction has now become reality with the arrival of the "hoverboard". This is *"a brand name for a form of transport for one person, consisting of a small board on which the person balances, with two wheels and a motor"*[472]. What makes it outstanding is that the device runs slightly above ground. The technology is still in its infancy and uses liquid nitrogen. Once the liquid nitrogen is cooled (below -145 degrees Celsius/-230 degrees Fahrenheit), the device runs for a maximum of 30 minutes on superconductors that facilitate the hover functionality[473].

6.3. Advantages

We now know our options, but what are the advantages of micromobility over other forms of vehicles? There are a few obvious ones that I've already shared but let's take a moment to appreciate some of the most noteworthy benefits of mass utilization of micromobility in our cities. This list[474] is from "The Urban Mobility Observatory," an EU-funded body :

- **Lower Traffic Congestion:** Though traffic got worse after the COVID-19 pandemic, in many cities around the world it is still much better than it was before the pandemic hit worldwide. According to a 2022 report by TomTom, a distance of 10km (6.2 miles) can take approximately 36 min 20 sec to travel in London, 29 min 10 sec in Bengaluru, India, and 28 min 30 sec in Dublin, Ireland[475]. TomTom's report shows us that leading the world in time spent longest in traffic is Dublin, particularly during peak traffic periods[476]. This translated into a loss of 145 hours on average stuck in traffic over the year for Dublin commuters.
- **Lower GHG Emissions:** Micromobility vehicles are phenomenally eco-friendly. Consider this from the EU Urban Mobility Observatory (ELTIS): *"Since micro-mobility devices don't use engines and don't consume fuel, these devices don't release any harmful emissions. Ujet and Sustainability report that*

replacing 8% of road vehicles with electric vehicles can reduce emissions by 80% by 2050. That's a significant positive impact on the environment, especially in the long run. If more cities encourage the use of personal or shared micro-mobility devices, it may be possible to reduce or completely remove air pollution sooner rather than later."[477]

- **Drives Greener Urban Development:** As with commercial solutions, public infrastructure developments are often driven by demand. The more people use micromobility in their cities, the more demand municipalities and governments notice. This then drives development in these communities. We then start seeing the construction of bike paths, bike lanes, docking stations, and other features of city landscapes that make it more practical for this type of commuter to get around as well as commuters using other forms of transport[478].

- **Lower Cost of Living:** The cost to maintain your bike, scooter, e-bike, or e-moped are simply worlds apart from the costs of maintaining the average car, whether ICE or EV. This makes this category of vehicles pocket-friendly for everyone, including students and individuals trying to make ends meet in low or medium-income countries. With those savings, you can put that money towards other lifestyle choices and expenses like food for your family, savings, education for yourself or children, vacations, and much more.

- **Increase Equity in Accessing Mobility-related Advantages**[479]**:** When a job or university campus is prohibitively far, rent is sky-high and you can't always manage to drive or take a bus, options for commuting that micromobility present offer real leverage. A student or worker on a bike can get to class faster—or in as much time as their colleague driving a Maserati—because they don't have to deal with congestion. The spare time they have can also allow for them to be more productive if they so choose.

- **Personal Health:** There are many physical activities that we can engage in to improve or maintain our health like walking, running, cycling, high-intensity interval training, and strength training. Manual micromobility vehicles allow you to reap the benefits of cycling whilst having the option of seamlessly letting the activity fit into your lifestyle. Many people have misconceptions as to who can or should cycle. Here's an excerpt from a New York Times article that I found encouraging for new or aspiring cyclists:

"Cycling is a lot more forgiving of body type and age than running. The best cyclists going up hills are those with the best weight-to-strength ratio, which generally means being thin and strong. But heavier cyclists go faster downhill. And being light does not help much on flat roads.

James Hagberg, a kinesiology professor at the University of Maryland, explains that the difference between running on a flat road and cycling on a flat road is related to the movement of the athlete's center of gravity.

"In running, when you see someone who is overweight, they will be in trouble," Dr. Hagberg said. "The more you weigh, the more the center of gravity moves and the more energy it costs. But in cycling, there are different aerodynamics—your center of gravity is not moving up and down." The difference between cycling and running is like the difference between moving forward on a pogo stick and rolling along on wheels."
Gina Kolata. 2007[480]

6.4. Disadvantages

As with all things, there are of course disadvantages to utilizing micromobility options:

- **Safety:** If we are honest, there are some times when we are absentminded or simply not as alert as we normally would be. The phenomenon is called "white-line fever" or "highway hypnosis"[481]. Fortunately, if you're safely strapped in your car and driving at a reasonable speed, in the event of an accident, the probability is that you'll be fine. But what if we hit a cyclist in that split second? The likelihood is that they may sustain far more injuries than you and any of your passengers will, if any.
- Among the world's safest cities to bike in are: Hangzhou (China); Utrecht and Amsterdam (Netherlands); Münster and Bremen (Germany); Antwerp (Belgium); Copenhagen (Denmark); Malmö (Sweden), and Bern (Switzerland); as well as Strasbourg and Bordeaux (France)[482]. Other cities, even in the developed world aren't that safe for cyclists; these include New Orleans, Arizona[483] and New York City[484]. You can take all the precautions you want as a cyclist but it's the responsibility of everyone on our roads to look out for each other and ensure mutual safety.
- **Messy Cities:** With the mass adoption of micromobility at a personal or commercial fleet capacity, we should expect there to be more stationary vehicles on our curbs and other places. Apart from the visual noise this adds, it may present challenges for each member of society including pedestrians[485].
- **Lifespan:** We've all seen vehicles of poor quality on our roads; when a vehicle outlives its expected lifespan the question of the waste this contributes becomes an environmental and safety concern. Companies such as Voi are attempting to integrate advanced technology to increase the lifespan of such vehicles:

"For example, Voi's 4th generation vehicle, Voiager 4, used cellular IoT, powered by the Ericsson IoT Accelerator, to

increase motor performance by 35% and increase its lifespan to five years." Ericsson. (n.d.)[486]

- **Impracticality:** Some weather, the value of your cargo, and whether you want to drive may make it challenging to ride your vehicle in a given scenario. As we may all agree, in some cases, it's just far more practical to take a plane, a car, a car over a bike, or a walk instead of a cycle. In many cases, this is often a matter of personal preference and perspective.

Cycling Is a Lifestyle

"Biking is about rhythm and flow. It's the wind in your face and the challenge of hammering up a long hill. It's the reward at the top and the thrill of a high-speed descent. Biking lets you come alive in both body and spirit. After a while the bike disappears beneath you and you feel as if you're suspended in midair. "
— *Gary Klein*

Cycling is a passion for me. members of my family, and many people in my network enjoy it too. At the date of publishing this book, I've run 120 marathons in New York, Berlin, Boston, Chicago, London, Paris, Washington, and San Diego. Sometimes I perform very well in a race, and other times I'm just fortunate enough to complete it.

However, whether I am simply enjoying a casual bike ride, training, or running a marathon, one thing I pay attention to at all times is safety because my guiding principle is to not get injured. This especially concerns me at times when I worry about my wife and kids on the road.

I am fortunate to live in a part of the world where road safety is very high. When our son was born, we bought a cargo bike (known in Dutch as a "bakfiets"). Ulrike, my wife, happily ferried our son to kindergarten through all types of weather: Mobility Dutch style. This was not some arduous chore. This was just another—often enjoyable—part of our daily lives. Whether in that scenario or many others, we ensured we were alert and responsive to sudden events on the road, that our bikes were in the best condition, and that we were in good health to ride.

- *Michael*

7. Powered Light Vehicle (PLV)

Let's take a closer look at the vehicles that fall into this category from ITF of new mobility vehicles. These vehicles, as stated earlier in this chapter include throttled e-bikes, mopeds, e-mopeds, motorbikes, e-motorbikes, rickshaws, and e-rickshaws.

Other names for throttled e-bike include twist and go, low-speed, throttle-assisted electric bicycle, e-bike, and scooter-style electric bike[487]. The report defines this vehicle as a "*light bicycle-like two- or three-wheel vehicle able to operate with no pedal action solely on the impetus of a throttle-controlled electric motor. Maximum speed and electric motor power ratings vary according to applicable regulations. Throttled e-bikes able to travel above 25 km/h (the exact limit depends on local regulations) are often regulated as mopeds*"[488].

A "moped" is *"a lightweight, low-powered motorbike that can be pedaled"*[489]. An e-moped is the electric version of the moped. A motorbike, another type of powered light mobility vehicle is defined as a street vehicle with two or three wheels, a seat, and the capability of reaching speeds beyond 45 kilometers per hour[490]. E-motorcycles are the electric versions of motorbikes. E-Rickshaws, also known as tuk-tuks or auto rickshaws, have a small motor, typically three wheels, and can carry both or either people or goods[491]. Similar to "e-motorcycles" or "e-" *anything* for that matter, "e-rickshaws" are the electric versions of rickshaws.

7.1. Advantages

Most of the advantages that we enjoy with micromobility can also be enjoyed with PLVs, with an obvious decrease in the health benefits you can personally reap. If your PLV is electric, it also provides the green benefits akin to that of micromobility.

7.2. Disadvantages

As with the advantages of PLVs, the disadvantages are also similar. It may be impractical to travel on a PLV in the dead of night in dangerous neighborhood, hostile weather, expensive and unprotected cargo, or a myriad of other negative factors that you may need to commute through. In many cities and locations around the world, infrastructure inadequacies will also affect the practicality of using electronic PLVs. If you know you have no stations to power your vehicle, the likelihood is you may only use it within cities where you are not limited by this constraint, or boycott the purchase of it entirely.

In those instances, a different type of vehicle would be safer and more practical. However, if there is something we can learn from communities in developing or middle-income countries in Asia or Africa, like Vietnam and Nigeria, it is that even PLVs and micromobility vehicles are still capable of bearing cargo with significant weight, including the rider, over long distances.

However, if your health situation, weather, or express needs do not permit, then other types of eco-friendly transport are worthy alternatives. That's where e-cars and similar vehicles can bridge the gap. With intermodal applications, we also have the opportunity to "connect the dots" of our transportation needs.

8. Intermodal Applications

Mobility of the future may often mean we have to utilize more than one form of travel to get to our final destination at a local or national level. Outside of a local or national level, this is similar to international travel, if you think about it. You're driven to the airport, hop on a plane, land at your destination, and take a car, van, or other mode of transport to your intended destination. These services typically allow users like you and me to access their services for interconnection of transportation through their apps.

Jelbi[492], a Berlin-based company established in 2019 is one example of this type of service that brings everything together. Registered users can connect to both public and private mobility services through the app. If you would like to move from point A to B on a bus, and to point C via train and finish your journey to point D on an e-moped, e-scooter, or taxi, this app can help make that happen when you type in your requirements. The service can present you with the best route for your journey and services available to you once you enter your destination and other details.

Jelbi is not the only service facilitating intermodal mobility services in the world. Other service providers have sprouted around the globe to meet the gap in the market, including Tracsis, HandyTicket, Kentkart, Moovel, HaCan, AN VUI, OpenMove, r2p, Cubic and Ubitransport[493].

9. More Innovation

Whatever solutions we are part of driving, as stakeholders we need to take careful thought about the whole ecosystem and not just the direct impact that these solutions have on us and our households. Our concerns ought not to end at its costs us or how it makes us feel. We need to consider the impact these innovations will have on our ecosystem long after we are gone and your children are left with a world that they inherited from us. A broader perspective is the only sustainable perspective that will serve us well.

More innovations are being born, including flying cars, as I stated at the beginning of the chapter. There's also the advent of artificial intelligence (AI) and how it will disrupt the mobility sector and impact our lives as we know it. But here, we will look at two innovations: ammonia-fueled cars and flying cars, which may sound like fake news…

9.1. Flying Cars

Electrically-powered vertical-take-off-and-landing aircraft (EVTOL) or flying cars are dual-mode (road and air) vehicles in which the driver has the option of taking to the air when desired or necessary[494].

The technology is still under active refinement but may soon be available mainstream, according to one company that's produced a flying car they dubbed the Helix[495]. Another company, Alef Aeronautics, recently announced similar news:

> "Human innovation has brought us to space and given us marvels like the Internet and artificial intelligence (AI), but as of 2023, flying cars remain largely the stuff of science fiction. That said, US start-up Alef Aeronautics is tentatively hoping for a different picture next year.
> The California-based company unveiled its Model A flying car prototype at the ongoing Detroit Auto Show while also confirming that the craft has been approved for test flights by the Federal Aviation Administration. So far the firm has not flown the vehicle for the public nor has it released footage of the car in flight." EuroNews.Net.2023[496]

9.2. Ammonia Engines

ICE engines designed to utilize ammonia as fuel offer another promising solution in the transportation sector:

> "Nowadays, ammonia is considered the most promising solution for abating GHG emission in large-bore internal combustion engines for marine transportation and power generation sector (Kurien and Mittal, 2022), where the low energy density characteristic of the current battery technologies makes electric propulsion unfeasible."[497]

Earlier this year, the Chinese company Guangzhou Automobile Group Co announced that they had developed the world's first engine that runs on ammonia[498]. This innovation is in collaboration with Toyota[499]. We can expect more developments in the months and years ahead.

10. Recommendations

We've discussed a lot regarding present-day and future mobility. Perhaps the question remains: What changes in our mobility can we each make to have a smaller carbon footprint starting today? Here are the efforts I recommend:

10.1. Changed Mindset

Changing our mindset of ideal modes of transportation, comfort, and lifestyle is part of what will enable us to live more eco-friendly lives as commuters, riders, and drivers. We need to learn that change in the transportation sector is not only inevitable: it's already started. For the sake of our future and that of generations to come, the decisions we make in our mobility will have a lasting impact.

10.2. Consider Total Cost of Ownership

Allow the total cost of ownership to help you determine the most savvy financial decision for you. If you count the cost of the environmental and financial impact of owning a fossil-fuel-guzzling vehicle, you may determine those are reasons enough to make drastic changes to the way you get around. Additionally, let's consider the costs over the life cycle. An emission free car might be more expensive at the time of the purchase but saves a lot of money over the life cycle through much lower running costs.

10.3. Shift to an Ecosystem Focus

In line with changing our mindsets, we need to shift the focus from ourselves as individuals to that of the ecosystem. Our decisions regarding the way we move need to be more aligned with the ecosystems that support it.

With such a shift, we start to make different, and typically better decisions about our movements. We begin to realize and operate on the understanding that not every trip that you and I make needs to be a car ride. We can derive great benefit from opting to walk or bike when factors allow.

10.4. Carefully Plan Trips

Whatever mode of transportation you choose to make on your journey, the most important part of it is to plan. If we focus on planning our mobility more, we may find we need to go on one or two shopping trips or other journeys when we currently have far more.

Taking these efforts, especially when it connects with the intermodal application of mobility, we solve our mobility needs that are complete, efficient, and more eco-friendly.

10.5. Use and Invest in Renewable-energy-powered Mobility

Many countries around the world, such as Norway, Iceland, the Netherlands, and China are making impressive strides when it comes to EV adoption[500]. Though Africa is the second-most-populous region in the world after Asia[501], it significantly lags in terms of EV adoption, especially in light of Africa having the highest urbanization rate of 4% when the rest of the world's rate stands at 2%[502]. In Africa, Rwanda is an outstanding example.

Despite EV adoption rates not being extensive yet, their efforts to locally manufacture their own electric vehicles puts them in the lead across the continent and other developing regions of the world[503]. Companies such as Ampersand, an electric motorbike taxi company, are not only making a positive environmental impact but helping their motorbike drivers increase take-home profits by 41%, a significant financial advantage over their counterparts who offer the same service using diesel- or petrol-fueled bikes[504]. When it comes to EV ownership, companies like Kabisa give Rwandans the opportunity and convenience of purchasing their EV from them, whilst benefiting from the government's incentive of a tax exemption of over 48% (value added tax, import duty and withholding tax)[505].

In one of the world's poorest countries, Malawi, we see glimmers of hope around steps in the right direction. Sky Energy Africa, a startup whose founder was listed as a Forbes Africa "30 Under 30" in 2019 makes available several electric cars and chargers for purchase, including Citroen Ë-C3 Electric Car, TATA Tiago EV, Wuling Air EV, Tesla Model X, and the BYD Tang EV[506]. Total Energies Marketing Malawi, another company in this small landlocked nation launched three electric motorcycle charging points in the capital city, Lilongwe on the 6th of December 2023[507].

10.6. Use Shared Transportation

Carpooling with neighbors to drop off children to school or to get to work is something most of us can start doing today. Using this and other types of shared mobility services is a legitimate way for us to move. If we use shared transportation, we cut our own financial costs associated with owning a car. If you live in a congested city, the likelihood is your car is idle most of the time anyway, and using shared transportation options is the smarter, cheaper, and more environmentally friendly option.

10.7. Intermodal applications may bridge the mobility gaps.

With a combination of the types of mobility services and mobility vehicles that we have access to, intermodal applications can make the process of getting from A to Z easier. For many of us, intermodal application to our mobility is not a new concept. If you are a business person or investor, this is an area of business worth investigating further. With what is forecast in the mobility sector over the coming years, this may be a source of great fortune for those interested in applying technological solutions to bridging any gaps in mobility.

10.8. Safety and Rules Will be Paramount for Mobility

As autonomous vehicles, flying cars, and other mobility vehicles enter commercial markets in a few years, the only thing that will make their convenience and existence positive for all of us is if we all follow the rules and regulations that keep us all safe as we travel.

These new types of transportation and even more conventional transportation rely heavily on our ability as commuters and drivers to look out for ourselves and one another.

10.9. Use or Lobby for Green-Sources of Energy

Sometimes, the electricity we use is generated through the burning of fossil fuels. This negates any "green" benefits because of the GHGs that are still released. Going green means abstaining from the use of fossil fuels in the back-end. Ammonia engines in the power sector may offer a viable alternative to coal as a source of energy. This is an additional matter we can lobby for if we live in parts of the world where coal generates our power.

10.10. Lobby for Change in the Transportation Sector

I've touched upon the power we consumers have that can influence businesses to make more changes, including environmentally friendly ones. The more we contact our favorite brands and enterprises with our requests and demands, the more they will realize what value addition matters to us and what they need to change to retain, or make more profit from, us as consumers.

Our lobbying should not stop with enterprises. We need to ensure that authorities and governments—wherever we are—know that the impact of mobility on transportation within their boundaries is not only a concern for us but one which we need them to address effectively for them to secure our votes and ongoing support.

11. Conclusion

In this chapter, we focused on aspects of present-day mobility and that of the future. We investigated the impact of mobility on CO2 emissions. Then we explored some of the health risks and problems that GHGs pose for us all.

Thankfully, extensive innovation within the transportation industry is helping to address these problems. The offer of emission-free mobility is increasing every day. Automotive companies have invested billions of dollars into emission-free mobility. But it starts with us; as consumers, we must demand change from all stakeholders and challenge our behavior. We can optimize our footprint by optimizing our way of going from A to B. This requires planning and careful execution.

Going by bike is often much better in the cities than by car. We cannot park cars anymore in the big cities. We will see more and more cities blocking cars and trucks.

Sharing cars with other people can be a practical and financially savvy lifestyle decision that improves our utilization of assets. Soon, autonomous cars will improve the utilization and safety of mobility vehicles.

In Chapters 2 to 4, we looked at how we live, what we eat and how we get around. In each chapter, I offered some recommendations for us to secure a climate-friendly future. In the next chapter, we are switching gears and focusing on the different stakeholders that we need to play a part in addressing climate change.

Chapter 5
WHERE DO WE NEED HELP?

> *"Incentive structures work, so you have to be very careful of what you incent people to do, because various incentive structures create all sorts of consequences that you can't anticipate."*
> Steve Jobs[508]

The enormity of the climate change problem means that we can't just wait for the government, non-governmental organizations, and influential people to take action. We all must take part. Throughout the book, we've looked at the facts, how we live, what we eat, how we grow, what we eat, and how we get around. We know that we all contribute to the severity of the problem—granted, at varying degrees. Now, we need to make a collective effort to fix it.

In this chapter, we'll look at where we need help to address our problems. We will look at economic systems, politics, ecological systems, and our social groups. Before we begin by looking at our economic systems, let's come to an understanding of what we mean by "WE", when we say "Where We Need Help."

"We" are all part of a society, which is *"a community, nation, or broad grouping of people having common traditions, institutions, and collective activities and interests"*[509]. At the most basic level of society, there is the family structure. We need to understand the connection between our societies (and the groups that comprise them) and the economic systems that they operate in.

1. Economic Systems

For us to understand "economic systems," we first have to agree on what an "economy" is. It is *"a complex system of interrelated production, consumption, and exchange activities that ultimately determines how resources are allocated among all the participants. The production, consumption, and distribution of goods and services combine to fulfill the needs of those living and operating within the economy."*[510] I was born and raised in Germany and therefore lived and operated within Germany's economy. Now that I have spent more time in The Netherlands, I've directly impacted the economy of this beautiful nation. If you work remotely or across borders, you can impact your local economy and the economy in which your work is produced or distributed.

An economic system is therefore the way authorities in a society allocate, organize, and appropriate resources within their jurisdiction[511]. There are several types of economic systems but capitalism and socialism are the most prominent. Most of us live in capitalist societies, and as such, much of what we own is produced or acquired through a profit and purchase basis, and typically is solely ours.

Economies are important for our survival as individuals, communities, and societies at large. Our societies would not survive without the economies that they operate in:[512]

> *"Economy, therefore, is a component of society; and society is the framework within which economy functions. Because of this relationship, every society has its own economy, and every economy reflects the needs and cultural attributes of society, as well as the major traits of the civilization in which it lives."*[513]

Our societies also have cultures. A "culture" is "the customary beliefs, social forms, and material traits of a racial, religious, or social group"[514]. I bring this up because the societies that you and I are a part of, the cultures that they uphold, and the economies that keep them running smoothly all contribute towards the health and detriment of our global climate. We all play individual roles in our respective societies, both officially and unofficially. Some of us have far more influence than others and can therefore influence our families or other groups in our societies towards the change that our climate desperately needs.

As we saw in Chapter 2, "Living of the Future," research shows that GDP has a positive correlation with global carbon emissions: "*If GDP increases one percent, CO2 emissions increase by 0.02 percent, assuming other factors are constant.*"[515] This clearly tells us that our economies are negatively impacting our climate.

As a result of our differences and what we each contribute to the economies we operate in, our interests and needs vary. On a grander scale, our interests and needs affect our respective carbon footprints. Common interests that you and I may have include economics, social change, work, time, environment, physical health, safety and security, and mental health[516]. Just as with individuals, organizations (businesses, governmental and not-for-profits) also have their interests—both private and public.

2. Incentivization

Whatever meets our interests or promises to do so, can be an incentive that moves us towards the agendas we want to see realized. When it comes to climate change, stakeholders may need various incentives to encourage the actions and behaviors we all need. Many organizations and authorities have made great efforts to figure out incentives that can work for the various stakeholders in our global societies towards addressing this challenge. However, there is always more room to brainstorm, collaborate, consult, bounce off ideas and work together at implementing solutions. We need to accept that not everyone will take positive action from an ethical or environmentally-friendly stance. Some influential stakeholders of our society will only take action when compelled to do so. That's where incentives come in.

2.1. Designing Incentives

Brookings offers a checklist[517] to refer to when designing incentives. I've adapted it below for our climate-related purposes:

- **Do the incentives target the right stakeholders?:** Will the targets of these incentives offer positive multiplier effects? If the incentives are geared towards a registered organization (e.g., a business), are they officially registered? Granting this incentive to unregistered businesses that are not local can take away benefits and even job opportunities from those that are registered.
- **Do the incentives target the right areas?:** Incentives may not make a big enough impact in lobbying households in The Netherlands, Germany or Switzerland to adopt solar energy at a household level. Global efforts directed at parts of the world that are lagging may have a greater impact.
- **Are they the right incentives?:** If cash incentives are offered, is the cash distributed upfront? Are organizations given incentives to hire local talent and skills? Are the incentives independent from any other financial responsibilities authorities have committed to distribute, or will these new incentives take away from some other program that already has financial commitment?

Looking back at my 30-year-long business experience, incentives alone do not change things quickly. Often, you need "a carrot and stick"—an approach "characterized by the use of both reward and punishment to induce cooperation."[518]

Scholars (Geest & Dari-Mattiacci[519]) share seven fundamental differences between "the carrot" and "stick" in question:

- The carrot is utilized for compliance, whilst the stick is utilized as a response to a violation.
- Carrots motivate by rewarding, while sticks motivate only by threatening.
- Carrots result in transaction costs through compliance and stick in the event of a violation.
- With carrots, all non-monitored citizens are treated the same as monitored violators. With sticks: all non-monitored citizens are treated the same as monitored compliers.
- Carrots may come with risks for those being compliant, whilst sticks may come with risks for violators.
- Only carrots come with a reward component.
- Each "carrot" (individual and general) has its own distributional effect for those who are compliant, while sticks (both individual and general) have the same distributional effects for compliers.

Sometimes authorities can structure incentives that simply don't work or have unintended consequences. When I shared with you about my family's solar solution, I briefly touched on this story that can help us understand why we need to take considerable care in structuring incentives.

2.2. "Green" Carrots and "Green" Sticks

Buying and selling carbon credits, or running cap-and-trade are additional options available in parts of the world. For those of us living where these are not known or recognized, we need the help of lobbyists and climate-friendly politicians in instituting them as viable additions to our fight against climate change. A "carbon credit" is a permit that can be bought and sold that represents one metric tonne of emissions that a business is permitted to release[520]. "Cap-and-trade" is "a system that limits aggregate emissions from a group of emitters by setting a "cap" (limit) on maximum emissions and is characterized as a market-based policy to reduce overall emissions of pollutants and encourage business investment in fossil fuel alternatives and energy efficiency."[521] Carbon credits are made by governing bodies and issued to businesses within jurisdictions that operate under a cap-and-trade system[522].

Some financial experts making compelling points about adopting the use of "green" carrots and "green" sticks:

> *"Instead of using green sticks to force change, why don't we use green carrots to entice change? After all, these approaches are not mutually exclusive.*
>
> *One way to introduce green carrots is to create a market for royalties from R&D into renewable and sustainable energy. Both the oil and gas and mining industries are already among the top developers of green technology patents, yet monetizing this research is difficult. A company can either use the knowhow and roll out the technology in-house, or be stuck with it. By the way, the greening of so-called dirty industries has perhaps the greatest potential to counteract climate change.*
>
> *In the biotech space, companies have already specialized in financing intellectual property (IP) in return for a share of the revenues generated from the finished product. Why is there no such system in place for green technology development?*
>
> *Alternatively, we could support dedicated royalty companies in the green technology space to open a new market. Investors could then invest in the shares of these green tech royalty companies and earn a profit from changing the world instead of saving taxes on burning it.*
>
> *We could even go a step further and learn from successful venture capital (VC) models in countries like Israel. Today, Israel is one of the world's leading tech hubs and much of the*

credit goes to the government-funded business incubator Yozma. In 1993, the government established Yozma by seeding it with $100 million in capital. Yozma supported early-stage ventures in exchange for a stake in the projects of up to 40% — provided private investors financed the rest. After seven years, the investors could pay back the government support from Yozma at face value plus interest. It worked, and in 1998, the VC market in Israel grew large enough for Yozma to be privatized.

The effectiveness of providing a carrot for investments should not be underestimated. Today, Israel spends more on R&D as a share of GDP than any other nation and is second only to the United States in terms of venture capital investments relative to GDP. Israel used carrots to transform its rusty 1990s economy to a modern high-tech one." Falk & Klement .2021[523]

A thorough discussion on incentives deserves a series of books. Regardless of the depth of such a discussion, the conversation needs to make practical sense to make a real impact or else it stays theory. To further illustrate the practicality we require, I would like us to look at the mobility sector as an example.

A Look at Carrots and Sticks of the Mobility Sector

"Economics, when you strip away the guff and mathematical sophistry, is largely about incentives."
John Cassidy[524]

As an executive leader in the automotive industry, I draw a lot of my wisdom and examples from my experience there. For instance, electric vehicles: We can try to incentivize electric cars as an example, but this will not be sufficient. In this example, I believe a stick is also necessary.

Many cities are using this stick to make it difficult or impossible to drive ICE cars into the city. In some cases, legislation is made causing ICE cars to be far more expensive to run and keep, leading to drivers switching to electric vehicles or shared options.

One particular state that comes to mind in this regard is California. It is taking a leading role in the US when it comes to offering and implementing carrots and sticks in the mobility sector. Trucks are required to be zero-emissions in the ports, with incentives built into the legislation that encourage the purchase of the same.

While a "carrot and a stick" is the fastest way to change behaviors it will not be applied in all cases to improve our climate. But when we have incentives it is desired to have them for a longer period so that consumers and companies can plan accordingly.

A good and bad example in this context is the incentive to buy electric cars in Germany. While the introduction was helpful to buy more expensive electric cars, it was taken away by the current government a few days before Christmas of 2023. This does not help the transformation to electric cars, as it irritates customers when they make long-lasting decisions.

Incentives can help only at the beginning. The Chinese government is very successful in supporting the transition to electric cars. This is very helpful at the beginning, but ultimately consumers decide, and they should decide on climate-friendly products and services without the benefits of incentives.

But as we have to change things quickly the role of simple, powerful, and effective incentives should not be *underestimated. We need them and we need to focus on the future and not subsidize the products of the past.*

- Michael

Whatever incentives are introduced, we need to make sure that apart from them being effective, they are not abused.

3. Politics

Whatever society you live and operate in, there is a head of state, be it a president, chancellor or royalty. Our leaders, regulations and policies that they institute are part of the politics that govern us and have power over us[525].

For those of us who are lucky enough to have a say regarding the leaders we need, we hope to see some of what we vote for come to fruition. Unfortunately, many of us around the world have leaders we vote for that do not take the actions they promised as they campaigned. Regardless of whether or not we can vote them out, there are millions of us around the world who cannot vote out leaders that fail us.

The most conflicting situation, I hope you would agree, is leaders and administrations that do a great job in several areas and yet fail us when it comes to climate change issues or even deny their very existence. Whether or not nations are taking action to fight climate change is also dependent on the policies that have been instituted in those nations. One relevant policy worth mentioning at this point is the "carbon tax" policy.

3.1. Carbon Tax

Carbon taxes are a type of tax that acts as a penalty for businesses for excessive GHG emissions[526]. They are a kind of carbon pricing, as are the carbon credits and cap-and-trade system (emissions trading system) that we talked about under Economic Systems[527].

Carbon tax differs from emissions trading systems in that "The emission reduction outcome of a carbon tax is not pre-defined but the carbon price is."[528]

Countries that have carbon taxes include China, Argentina, Chile, Colombia, Japan, Kazakhstan, Korea, Mexico, New Zealand, Singapore, South Africa, Sweden, Ukraine[529], Finland, Norway, Sweden, Denmark, the Netherlands, Germany, the United Kingdom, Australia, select places in the United States (such as Boulder in Colorado and Montgomery County in Maryland), and some provinces in Canada (Alberta and Quebec)[530].

According to the IMF, carbon pricing demonstrates great promise for fighting our climate change problem as it can readily be implemented, particularly in mitigation strategies; it is gaining momentum, and can be steered at a global level [531]. The IMF adds that a thoughtfully engineered price floor (legal minimum prices) may be more effective than alternative options[532].

4. Ecological

In the chapter on food and food systems, we discussed types of agricultural practices and took time to understand the benefits of regenerative agriculture including agroforestry and no-till agriculture. Certain types of effects of climate change are ecological and could be added to the mix with the help of subject matter experts from indigenous groups and academia alike.

We need to know how we as individuals can protect the natural environment, beginning with local efforts. With rising temperatures destroying the diversity of organisms, sea levels eradicating existing coastal systems, and melting ice causing the displacement of living organisms in those regions[533] and more, we are left with the question of how to help correct this or assist in adaptation. Wildlife is already dying out and becoming extinct because of our behaviors. With further disruption of their homes or conditions that make it conducive for their survivor, it is simply a matter of time until these animals and plants are known only in books and digital format.

4.1. Define Wildlife Areas

As animals cannot advocate for themselves, it is our responsibility. Some cities and towns need wildlife areas defined. Other areas need wildlife corridors created to help animals migrate from one area to another. Wildlife corridors are *"connections across the landscape that link up areas of habitat. They support natural processes that occur in a healthy environment, including the movement of species to find resources, such as food and water.*

> *Corridors can contribute to the resilience of the landscape in a changing climate and help to reduce future greenhouse gas emissions by storing carbon in native vegetation. They can also support multiple land uses such as conservation, farming and forestry."* Australian Government.[534]

Local communities can create these but authorities are taxed with the bigger picture of creating more significant corridors. Wildlife corridors of a larger scale may need to span up to hundreds of kilometres[535].

5. The Experts

For us to understand climate change, we need access to current and accurate big data. Most of us would struggle to interpret that data without the help of qualified individuals. As climate change is a complex problem, it needs efforts across disciplines. As this is a collective problem, we need the brightest and best of us across fields, to contribute towards fixing it. As some sociologists aptly put it:

"Climate change will be one of the greatest transformational forces of the twenty-first century...Scholars working on consumption behaviour have been challenged to focus their attention on the actions that have the greatest consequences for climate change (Stern & Wolske 2017). There is good evidence that household actions reducing direct energy consumption, encouraging changes in the supply chain of consumer products, and engaging in distributed production of renewable energy can help reduce climate risk. But some of those actions are far more consequential than others. It behoves sociologists working on consumption to focus attention on high-consequence actions, such as home weatherization, transportation, and food choices rather than routine practices that have less impact. And it is crucial to consider the supply chain of production, consumption, and waste disposal (Stern & Dietz 2020). The same logic can be applied to the study of citizen decisions. The literature has focused on climate beliefs, risk perceptions, and to some extent general policy support. But there has been relatively little attention to how these variables influence policy, or to what drives actions like voting, donating to movements, and otherwise engaging in political change. While the argument to focus our research efforts on what matters has mostly emerged in the micro-scale literature, the argument is certainly salient to meso- and macro-scale work as well.
Spillovers, in which one decision either increases or decreases the probability of another, may substantially change the dynamics of climate action. At the micro-scale, modest initial consumer actions might block larger and more consequential consumer actions and political action. But it can also be argued that easy actions are the first steps in making commitments

and developing an identity congruent with both consumer and political action. Evidence and theory about spillovers are accumulating and remain mixed, with some evidence for both positive and negative spillovers (Truelove et al. 2014, York 2017). At the macro-scale, several studies have raised concerns that the adoption of renewable energies and energy efficiencies does not fully displace fossil fuel use but rather contributes somewhat to increased energy use." Dietz, Shwom & Whitley. 2020[536]

6. Our Social Groups

Within our families and our communities (offline and online), we hold influence to varying degrees. It is easy to overlook it and even think that we hold no power, but that could not be further from the truth. Let me share with you a very personal example of my own about just that.

As Ladies Lead

"Women, whether subtly or vociferously, have always been a tremendous power in the destiny of the world."
— Eleanor Roosevelt, "It's up to the Women"

If you've read this far, then you know that almost my whole family eats a plant-based diet. As you may also recall it was my daughter and Ulrike that got us started on this lifestyle. We have a dual-income household, and my income is primary but when it comes to influence, and how my wife and daughter managed to shape the way we live, I am grateful for their influence and impact.

If it was not for them, I may not have arrived at my current passion for the environment and for healing our planet from the challenges of climate change. I know that it was their concern about eating healthier, their impact on the meals we eat, and even their preparation of them daily that firmly established our dietary lifestyle change. They spent several years eating a plant-based diet before I eventually started.

With this as a basis, my curiosity for the larger picture beyond food grew. This phenomenon of the power of women in the household is not unique to my household. Consider the influence your maternal figures had in your life. Did they shape your perception of life and even the lifestyle decisions you make? I may be a top executive in global business, but in my home, influence is deeper and diverse—most likely as it is in your home too.

- Michael

Even if our families do not provide the support we would like to advocate fighting the effects of climate change, there are numerous online communities—many of them free—that give us the community and access to join the collective efforts we desire to make. It boils down to being willing to research and choose the activities we take part in.

7. Recommendations

We need a lot of help to curb climate change. As such, these recommendations are purely a sample of those that I felt necessary to highlight:

7.1. Everything Needs to Have a Price

If you live in a region of the world where businesses are not taxed for their carbon emissions, you can advocate for this development at a policy level with your local authorities and environmental organizations. It should cost businesses for the emissions they contribute towards harming our planet because if it doesn't cost them financially, it will end up costing us all environmentally. Similarly, it is only rational to increase the price of meat (and other products that cause significant emissions) to offset the negative impact that those commodities have on our planet.

7.2. Keep Things Simple

Regardless of where you are, legislation needs to be so simple that any adult can understand and interpret it for themselves. In many countries, legislation and policies—particularly on tax—can be excessively complicated, leading regular citizens to require the aid of finance experts to make sense of it all. If we want to help heal our planet, we need simple VAT schemes which support climate-friendly products, services, and solutions.

7.3. Help From Experts

Help from sociologists, economists and other experts in addressing the negative effects of mass adoption of solar energy and other types of sustainable energy solutions:

> *"At the macro-scale, a number of studies have raised concerns that the adoption of renewable energies and energy efficiencies does not fully displace fossil fuel use but rather contributes somewhat to increased energy use. As a result, emission reductions are less than might be expected from an analysis that considers only technology (York 2012, York & Bell 2019, York & McGee 2017). It is clear that a more integrated theory of decision making that takes account of spillovers is needed." Dietz, Shwom & Whitley. 2020*[537]

7.4. Incentivize Key Stakeholders

Some stakeholders find being part of the solution a reward in itself. Others need more. We need to make the effort to provide those incentives where philanthropy is deemed insufficient for various players in overcoming climate change. We need effective incentives for all stakeholders, particularly big businesses and consumers. When it comes to big business, tax incentives (tax exemptions, tax reductions, tax refunds, and tax credits) are all elements that could help big business do better[538]. Another type of incentive that may work well with key stakeholders, and particularly businesses is subsidies, which are "grants, or sums of money, that governments give firms in an effort to boost business."[539]

Another type of incentive that can help us is charging businesses more for waste. If we make waste management expensive, businesses (particularly the larger polluters) will be forced to make more environmentally-friendly choices or be forced to shut their doors.

7.5. Teach Another

Teach someone what you learned in this book. Begin at home, with your family and friends. Let your family be the model.

You can influence your families and other communities that you are a part of, whoever you are. You may even have more influence in your sphere of influence and culture because of the work you do or what you symbolize in your society. If you are a politician, teacher, artist, musician, community leader or celebrity, this may be easy for you to harness.

If you don't know where to start, consider starting by sharing your concern about the change in local weather and the behaviors of your local community in person and on your social media accounts. Help others make the connection between whatever unique behaviors your community does that negatively impact the environment such as cutting down forests, driving ICE vehicles, burning fossil fuels for heating or other uses, eating meat, and building with cement or steel.

Take time to share the language around climate change, as having the right words is the beginning of coming to a collective understanding[540] that leads to collective action. Referring to the harm we cause our planet as "toxic violence" is an example of the type of language we need to normalize that may help us make the mental shift that we need as we address climate change.

7.6. Lobby our Leaders

Leaders, particularly those we elect, need our votes and are often willing to champion our concerns if they receive the opportunity to lead us. As leaders around the world continue to see the devastation caused by climate change, we have the opportunity to direct their efforts as citizens that they govern. It is our duty as responsible citizens to bring important issues like that of climate change to our leadership's attention and request that they represent us and take the actions we need to see realized.

8. Conclusion

We all need help to fix this problem. If we look at it as individuals, addressing climate change is simply impossible. But we can do this together. As I close this chapter, consider this story:

Are We Going to Die?

"... I Guess I'm Going to Die..."
Colin Jost.[541]

There is no denying it: fighting climate change is the kind of problem that may overwhelm us at just the very thought of it.

Some of us may even be moved to inertia, as one comedian jokingly commented:

> "We don't really worry about climate change because it's too overwhelming and we're already in too deep. It's like if you owe your bookie $1,000, you're like, 'OK, I've got to pay this dude back.' But if you owe your bookie $1 million dollars, you're like, 'I guess I'm just going to die.' " Colin Jost, SNL[542][550]

The truth is though, the climate change problem is not just your problem and mine. It's a problem that we, and generations before us created that will impact us and the generations after us. Our sons and daughters will have to live with the unpredictable and inhabitable conditions we create simply because we failed to take adequate actions.

- Michael

The climate change problem is a pricey one for each of us, though to varying degrees:

> "We all will have to pay a price for burning fossil fuels, but unfortunately the bulk of that price will not be paid by those who burn fossil fuels. It is a classic problem of a negative externality: The profits of an activity — in this case, burning fossil fuels to generate energy—are privatized, while the costs, to human health and the environment, are socialized." Falk & Klement. 2021[543]

In this chapter, we looked at where and what we need help with, and who can help to address climate change. In the next and final chapter, we'll focus on what each of us can do as individuals.

Chapter 6
EVERYBODY CAN

"Be the change that you wish to see in the world."
— *Mahatma Gandhi*[544]

Here we are at the final chapter where we discuss the most practical of all topics titled "Everybody Can." We can't wait for our politicians, social activists, and celebrities to change the world for us. We also have power. We too can change the trajectory of our planet's health. Each of us can have an impact—both social and environmental.

Let's begin by taking a moment to reflect on what we are doing as individuals, then move on to measuring tangible progress, learning from each other, and what we can start doing right now. Here we will focus on the social positive impact on the environment that we can contribute to. Social impact is "a significant, positive change that addresses a pressing social challenge."[545]

Everything has a first step. It can be difficult but it can make a huge difference. With such a complicated problem, we need a framework or thoughtful approach to tackle it. One popular approach we can use is the theory of change, *"a roadmap or planning process that presents how change can be attained."*[546] This is a helpful framework for complex and long-term challenges[547]. What I know works well for me and my family is these deliberate, yet small, consistent and incremental steps. Through research, I learned that it sounds a lot like the nudge theory of change, a popular theory of change that involves setting ourselves up for success using nudges:

> *"The concept of nudge theory has its roots in behavioral science. It suggests using subtle direction to help steer people towards taking a specific action or making a decision through*

a voluntary mechanism rather than enforcement. Often Nudge theory is used in conjunction with other theories and models, particularly after an initiative has been implemented."
Johnson-Woods & Wardlaw[548]

As social beings, driven by our habits, generating more climate-friendly habits can do us good and save the planet. Consider the following story.

Going Up!

"A nail is driven out by another nail; habit is overcome by habit."
— Erasmus[549]

Forming new habits is hard. Switching to a fully plant-based diet was not something I managed overnight. Becoming fit enough to run 120 marathons—and counting—is an excruciating effort. The results were worth it though and I would do it again in a heartbeat. Over the years, as I did these things and other challenges, I learned that there are small steps I can take to generate habits that can help me get the results that I want. I gently "nudged" myself to where I needed to be. Sometimes these new habits were so simple that you could easily overlook them. This reminds me of research carried out by a group of social scientists over forty years ago in Canada at a university:

"They added a 16-second delay before the doors closed to see if more people would use the stairs. They did. No surprise there. But what was particularly interesting was that when the elevators returned to normal many people continued to use the stairs, cutting energy costs and raising students' fitness levels.

What's more, the elevator trick cut energy usage far more than when the university simply posted messages asking people to take the stairs. Adding tangible friction into our routines, in other words, can sometimes beneficially change habits... This has demonstrated how powerful oft-ignored cultural patterns are. But it has also shown these can change." Tett. 2021[550]

Consider how the small experiment above translated to a positive environmental impact through the saving of energy, and increased health benefits from students.

- Michael

If you still think your small steps and mine don't matter, then refer to the research. The problem with nudges is figuring out which ones will work for you and me as individuals[551]. However, once we figure out what does work, these small steps not only count but are making the change we want to see.

Here is another example.

Adapting Our Taste Buds

"The world's strongest animals are plant eaters. Gorillas, Buffaloes, Elephants and me."
Patrik Baboumian[552]

One of the hardest things to get used to once you adopt a plant-based diet is finding suitable animal-protein replacements. The texture, aroma and taste are so unique, and not the easiest to replicate in plant-based form. Thankfully, with an adventurous spirit, you can discover new recipes and alternatives like jackfruit and mushrooms that can bridge that taste gap. Establishing this change boiled down to dietary behavior change. One study found that on average we make 226.7 daily decisions about food alone[553].

One experiment funded by UNDP (United Nations Development Programme) conducted by UNDP Egypt Accelerator involving 230,000 all-women participants from an online group, identified five behavioral insights utilizing nudges to influence the participants towards more climate-friendly diets[554]. This experiment measured the following criteria: "1- diversity of recipes consumed, 2- convenience, in terms of time and accessibility, 3- the weight of nutritional value, and 4- the motive to be part of a large-scale impact."[555]. In their results[556], they found the following five behavioral insights that impacted the likelihood of a positive change in their diet:

- Family acceptance
- Accessibility of ingredients
- Easy recipes
- Price of ingredients
- Healthy replacements

When I saw this research and reflected on my household, it all made sense. Our daughter was the first in our household to go plant-based. I followed after stumbling across a plant-based diet challenge I decided to take on for 30 days. As soon as I took the challenge and saw how doable it was to keep it going, the long-term choice became easier. As we are fortunate to have access to locally grown and fresh fruits and vegetables, enjoy making meals, and have access to healthy replacements, making the lifestyle change was easier than we initially thought.

- Ulrike

When it comes to subtly guiding ourselves and each other towards better care of our environment, we need to be very careful because sometimes, we can take a nudge but feel unmotivated to make more impactful efforts: Researchers studying nudges over 4,000 participants found that people who felt they had taken a small action, like subscribing to a program supporting climate change, would decline supporting carbon tax (which is even more impactful)[557]. One of the scientists involved in the program stated the following:

> "The reticence of the first groups may be explained by a phenomenon known as 'single action bias,' which is when you feel that you've done enough to address a particular issue, so you decline to do more... Nudges should be used as complements to standard economic policies. It turns out that when we tell people that nudges aren't that effective [on their own], compared with carbon taxes, and when we tell them carbon taxes can lead to improvements in other areas of their lives, then we find this effect disappears," she [Emily Ho] said. Fordham News, 2019[558]

For example, eating a plant-based diet is not enough, just as planting a few trees each year is good but also insufficient. Research shows that if each person planted approximately 160 trees, we would cancel out the global CO2 emissions of one decade[559]. However, the likelihood of each of us planting that many trees may be low. Still, the fact that deforestation is happening at an area roughly equal to 27 soccer fields per minute[560] should make us think about making extensive tree planting part of our individual and global lifestyles.

Lao Tzu, the Chinese philosopher once said: *"Watch your thoughts, they become your words; watch your words, they become your actions; watch your actions, they become your habits; watch your habits, they become your character; watch your character, it becomes your destiny."*[561] Instead of *your destiny*, I would say *who you are*. If we are actively doing nothing to stop climate change, we are the villains of this climate change story. If we decide to take positive actions and are determined by them, then we become the heroes or protagonists. And if we all join hands to save our planet, then we are a force that will change not only our planet but also positively help generations to come. Each of us needs to decide: am I the villain or the hero when it comes to climate change?

1. What Are We Doing?

If we want to effect change, we will rely on our abilities to have a social impact on our communities. One expert on the matter, and former lecturer of applied sciences, Amsterdam, Marian Spier, shares that for there to be a social impact you and I need the following five things: opportunity (the chance), strategy (clarity), passion (craze), the right mindset (courage), and purposeful determination (considerate and determined)[562].

Climate change has created "the chance" or opportunity. The future of generations to come should give each of us the passion, and purposeful determination, and also equip us with the right mindset. If this was a problem that only one person could solve, we could delegate all efforts to them, and applaud them for their success. This is also a problem that needs addressing immediately and will likely show its biggest rewards in the long term, perhaps when many of us are gone and dead. However, the opportunity or chance to create an impact is for everyone, starting right now. There are those people who've already seen this opportunity and have run with it. Some of us are still catching up. There are whole industries and generations of wealth being created by those who manage to see the income generation opportunities that creating sustainable products, services, and solutions offers. These people are disrupting various sectors and changing the way we live for the better.

In such a practical chapter, I've started in quite a general way. You might be thinking, "Enough talking, what are *you* doing right now?". That's a great question. Let me be transparent with you.

What Am I Doing?

"Ideas alone will not change the world, but deeds will."
— *Dax Bamania*[563]

Here is what my family and I are doing to fight climate change:

- *Committed to a plant-based diet since 2020*
- *Buying local fruits and vegetables*
- *Driving electric cars*
- *Cycling to the local stores for groceries (whatever the weather)*
- *Walking and running when we can*
- *Powering our vehicles and homes with our photovoltaic panels*

Utilizing public transport, particularly when we travel further afield.

- *Michael*

2. Measuring Tangible Progress

It is all well and good to toot one's own horn but without measuring those efforts you have no idea how well you are doing or what you can do better.

I Can Do Better. Can You?

"What gets measured gets done, what gets measured and fed back gets done well.."
- *John E. Jones*[564]

I shared with you what my family and I are doing to be transparent and hold myself accountable. This information would not be complete without me sharing my carbon footprint.

Though I cannot vouch for the accuracy of the free online service offered by WWF (https://footprint.wwf.org.uk[565]) I resorted to using it last January 2023. According to its calculations, my carbon footprint is 9.773 tonnes of CO2. The German average is 12.37 tonnes and the global average is 4.7 tonnes of CO2[566] (according to the International Energy Agency in 2023).

In comparison to the global average, despite my existing efforts, my carbon footprint is much higher than I would have liked. To help myself understand this further, I used the US Energy Protection Agency's (EPA's) "Greenhouse Gas Equivalencies Calculator"[567]. This is another free online resource that helps you 'translate' emissions into concrete terms so that they are easier to understand and less abstract.

According to EPA's GHGs Equivalency Calculator, my carbon footprint of 9.77 metric tons is equivalent to GHGs from:

- *2.2 gasoline-powered passenger vehicles driven for one year.*
- *25,046 miles driven by an average gasoline-powered passenger vehicle.*

It is also equivalent to CO2 from:

- *1,099 gallons of gasoline combusted*
- *10,944 pounds (4.97 metric tonnes) of coal burned*
- *One year's electricity for 1.9 homes.*

With these helpful equivalencies the picture is far clearer and quite disappointing. To be frank, I assumed I was doing a lot better than I was, up until now. But now, I have a tangible goal. I will now aim to reduce my carbon footprint by approximately 40%.

- Michael

Do you know your carbon footprint? If you don't, stop reading right now. Go online and search for a reliable online service to quickly do an assessment. You could use what I used: https://footprint.wwf.org.uk. Within minutes you can know your carbon footprint. If your result feels too abstract for you to conceptualize and act upon, take the figure of your carbon footprint to the US EPA's GHGs equivalency calculator or similar resource, submit your data and see for yourself, what your carbon footprint translates to.

As we measure our health-related metrics to see opportunities for improvement, we create the likelihood of better health. Likewise, in caring for our planet, the least we can do is know our carbon footprint as the first step, before making actionable and real-life changes that have an environmental impact. Measuring your CO2 footprint is helpful and gives you a sense of self-awareness that you probably will not have any other way.

3. Learning From Each Other

It is easy to say, "Let's all start using renewable energy!" The cost of going fully green at a global scale (mentioned in "Living of the Future") is extensive with solar and wind facilities requiring more concrete, aluminium, iron, copper, and glass than fossil fuels or nuclear energy[568].

That means we need to figure out how we can neutralize that negative impact, which calls for collaboration from stakeholders in various fields in putting together reasonable solutions.

Another concern about going green may be the job losses in communities that depend heavily on the acquisition and refinement of fossil fuels. In such towns and cities, research and implementation of solutions on how to help these communities find alternative means of income is vital. Otherwise, how does someone choose renewable energy when they or their loved ones' whole livelihoods depend on fossil fuels? In such a context, these people are given impossible choices. As a global community, can we help them, their leadership and other stakeholders come up with workable solutions that make this shift feasible for such communities in every way, particularly financially?

As this is a complicated problem, we must accept that there is not just one solution to fixing it. Rather, a variety of actions, strategies, components, and solutions together make up all the pieces of a puzzle that we work collectively to assemble.

There are various online resources we can use to learn more and be more informed, many of which are free. As we saw with the carbon footprint web app and EPA's GHG equivalencies calculator, platforms exist that focus on climate change and can help us do better at being part of the solution.

I believe the key to learning, and applying that knowledge, is to make it easy, social, and fun. Spending time sharing what you've learned with a friend during a commercial break when your favorite game is on, or gifting your child a sustainably sourced treat or toy, and explaining to them what that means are all ways we can participate in helping others learn about climate change. Going on your accounts on social media and sharing with your network of friends and contacts are also simple, social and potentially fun additions to what you can do as well. How about sharing your carbon footprint with your friends and explaining what it means in terms of equivalencies? I am willing to bank on at least one person among them being interested in learning their own carbon footprint.

4. What Can We Start Doing Right Now?

The world does not need pledges: It needs action right now from each of us. With varying degrees of influence and impact, our contributions will differ, but they all matter. I shared with you my carbon footprint, and how I feel the need to do better and lower my carbon footprint by 40%. So here is what I plan to incorporate into my pre-existing list.

Changing the Future

"When you know better, you do better."
- Maya Angelou[569]

In addition to my existing efforts, I am committed to:

- **Do less international travel:**
  As an executive international businessman and athlete, the years have seen me all over the world. Where possible, I will skip the flight and simply conduct the meeting online.
- **Recycle more:**
  I shared how my wife's family are excellent at this. I know there is more opportunity for me and my household to recycle more, not just clothing and toys.
- **Live in houses, which have a much better CO2 footprint (and smaller):**
  Having lived in a solar-powered home, I wouldn't want to live any other way. That being said, I will make it

my mission to only live in homes that use renewable energy for the totality or bulk of their energy needs.
Use a heat pump: Using this heating system, solar panel system and air, we insulate our home.
- **Plant trees:**
  This could not be an easier positive contribution to the environment.

- Michael

As a result of the roles we play in society, we need to look not only at the direct impact we have on fixing this problem but how our influence and even power can help the greater good.

Whatever you and I decide to do needs to be planned and measured based on the extensive facts we have on the ground, and executed consistently. Our Earth is worth the effort.

5. Recommendations

This is the most practical of all my chapters. I encourage you to fill in the following information:

5.1. List down what you are already doing to help stop climate change.

I shared mine, now it's your turn. Let your list give you perspective.

_____

_____

_____

_____

_____

5.2. Find out what your carbon footprint is

Use the US EPA's GHG Equivalencies Calculator if that helps you understand your carbon footprint better.

5.3. Make a Carbon Footprint Goal

If you know what your carbon footprint amounts to, are you happy with it? How much less carbon emissions would you like to lower your carbon footprint?

_____

_____

5.4. Keep Learning and Teach Someone What You Have Learned

Some of the solutions we need can only be born through collaboration and partnership. We need to engage others and share what we know to make the differences we want to see.

5.5. Create Your 'Happy Climate List'

What are you willing to do differently? You can refer back to my list for inspiration if that helps. If you have ever taken on a discipline such as fasting for some time during Lent or Ramadan, or for whatever purpose you saw fit, committing to do it for some time can make the difference between getting started or never. Make it easy to commit to by choosing only 3–5 things and specifying the duration you are willing to do them for next to each activity on your list. For example: Only eating plant-based foods (activity) and doing it for the next 30 days (timeframe).

	Activity	Timeframe
1.	_____	_____
2.	_____	_____
3.	_____	_____
4.	_____	_____
5.	_____	_____

6. Conclusion

In this chapter, I hope you felt drawn to look at your own efforts towards making a happy planet. We began by looking at the concept of impact, particularly social impact concerning climate change. Subsequently, we looked at measuring results, learning from each other, what we can start doing, and recommendations. At this juncture, I hope it is clear that we need to accept a degree of discomfort and even sacrifice. If you have ever done something difficult that resulted in a reward such as fasting (abstaining from food or something else), you will know it is typically not a pleasant experience. It's painful. Some of the changes we will need to make will elicit a discomfort that could be called "pain". However, like with each of these, the benefits are worth it.

If you took the time to list your existing climate-friendly efforts, determine your carbon footprint and list your "Climate Happy List", well done. We are almost at the end of this book. Next, let's move on to my final words: The Conclusion.

CONCLUSION

"It's always something, to know you've done the most you could. But don't leave off hoping, or it's of no use doing anything. Hope, hope to the last."
— Charles Dickens[570]

We have discussed a great deal surrounding climate change. In Chapter One we began by focusing on the facts, what's being done and what adapting to or mitigating climate change may require from us. Chapter Two: We looked broadly at how we live with a focus on housing, housing materials, and their impact on climate change. In Chapter Three we focused on understanding food and food systems, and we considered what a more climate-friendly future would require from us. Food and food systems are integral to our existence and a significant factor impacting our climate.

Then in Chapter Four we zeroed in on mobility: the vehicles and services available now, and those that would benefit the environment by becoming further popularized.

We shifted gears in Chapter Five; we stopped looking at our impact from a causation perspective and looked at the various agents of change that can support a healthier planet. Here we looked at the problem from various angles: economic, political, and ecological.

In the final Chapter, we concluded with the most practical chapter of all: "Everybody Can". Here I drew your attention to what you and I can do as individuals, beginning with understanding what we are already doing, discovering our carbon footprint, and what we can commit to doing going forward. I offered as much transparency as I could, shared my commitment with you to lowering my carbon footprint, and challenged you to do the same immediately. From

Chapter Two to Chapter Six, we explored recommendations on how we could address this problem.

Now, here in the conclusion, I realize a stark reality. Often there is a divide between what we talk about, what we are doing, and what we would like for ourselves and future generations. We can talk about changing our habits and impact on the Earth, and fail to actualize it. I hope that having read this book I've accomplished two things:

- Helped you to see the urgency of the climate change problem.
- Helped spur you on to action that makes for a healthier climate.

You and I cannot live in harmony if we are destroying ourselves, others and the planet.

The Future

> *"It is quite obvious that the human race has made a queer mess of life on this planet. But as a people we probably harbor seeds of goodness that have lain for a long time waiting to sprout when conditions are right. Man's curiosity, his relentlessness, his inventiveness, his ingenuity have led him into deep trouble. We can only hope that these same traits will enable him to claw his way out."*
> — E.B. White[571]

When Ulrike and I were growing up in Germany—she in East Germany and I in West Germany—we had rich childhoods. When I say "rich", I don't mean financially. We both came from humble beginnings and can recall our parents doing all they could to lift us from the financial limitations they had. Our childhoods were rich because we were loved; we had the beautiful German countryside to explore, and to grow and learn in. I remember picking berries and mushrooms in spring, climbing trees, playing hide-and-seek in the woods, and stargazing on clear nights. Many years later, our two kids have enjoyed both the joy of the outdoors and the lifestyle we are lucky to afford.

My concern is now what my children's children will have, and what their children's children will have too. Will they get to run in open fields of dandelions and sunflowers? Will they be able to swim in streams and marvel at the myriad types of colorful butterflies that would grace their sights in the great outdoors? Will they know how wild berries like that of Walderdbeeren and Stachelbeeren taste?

I know that as things stand, almost all of this may not exist and will only be available as things I tell my grandkids about in stories from my past. Many things about Germany need to remain in the past, but these beautiful and marvelous gifts of nature and our heritage are not among them.

Climate change will rob us of our futures if we continue what we are doing now. In the Introduction, I referred to our planet as Mother Nature. Many of us have enjoyed her generosity for years. It's time for us to take care of her.

- Michael

We talked about a great many things throughout the book that can be attributed to human ingenuity. Many of them, such as internal combustion engines in Chapter Four, the factory systems, the power loom (used for mass production systems), the airplane, the telegraph and many other communication devices are inventions from the Industrial Revolution[572]. Additionally, the world benefited from new basic materials (steel and iron), and new energy sources (fuels and motive power such as coal, steam, petroleum and electricity)[573].

The Industrial Revolution changed our whole way of life as humanity. Now we need a new revolution. We need to change our whole way of life as humanity for the sake of our planet. If we do not, we must accept that our planet's conditions will change, and we may be forced to adapt to living conditions that are hostile, and abandon others when they become inhabitable.

We cannot afford to read this book and forget its contents. Staying idle is not an option. Half measures are not acceptable. Determining what you will do to stop climate change and getting started is the only way. You can do it. I can do it.

If history has taught us anything, it is this: people who set their sights on doing the extraordinary with passion and determination often accomplish it.

Be a powerful force. Change climate change now.

GLOSSARY

ADAS	Advanced Assistance Driver Systems
AI	Artificial Intelligence
AV	Autonomous Vehicle
BEV	Battery Electric Vehicles
CDC	Center for Disease Control (a US agency)
CMCC	Centro Euro-Mediterraneo sui Cambiamenti Climatici
EP	Environmentally Preferable
EIA	US Energy Information Association
EVTOL	Electrically-powered Vertical-take-off-and-landing Aircraft
FCEV	Fuel Cell Electric Vehicles
BIPV	Building-Integrated Photovoltaics
BTEX	Benzene, Toluene, Ethylbenzene and Xylene
EPA	Environmental Protection Agency
G20	Group of Twenty
GDP	Gross Domestic Product
GHGs	Greenhouse Gasses
HEV	Hybrid Electric Vehicle
ILO	International Labour Organization
IMF	International Monetary Fund
ITF	International Transport Forum
Km	Kilometers
LEED	Leadership in Energy and Environmental Design
LULUC	Land Use and Land-Use Change activities
MAAS	Mobility-as-a-Service

MHEV	Mild Hybrid Electric Vehicles
OECD	The Organisation for Economic Co-operation and Development
PHEV	Plug-in Hybrid Electric Vehicle
PLV	Powered Light Vehicle
PMT	Passenger Miles Traveled
SAE	Society of Automobile Engineers
SDG	Sustainable Development Goal
SMMT	Society of Motor Manufacturers and Traders
SNL	Saturday Night Live
TCO	Total Cost of Ownership
UN	United Nations
UNDP	United Nations Development Programme
UNECE	United Nations Economic Commission for Europe
UPF	Ultra-processed Food
US	The United States of America
V2G	Vehicle-to-grid (V2G) Technology
WFP	World Food Programme
WHO	World Health Organization
WWF	World Wide Fund for Nature

About the Authors

Co-authors Michael Lohscheller and Ulrike Louis debut their author journeys with Happy Climate—Happy Life. Michael is an international CEO with leadership experience of over 30 years on three continents in automotive industries. He is also an athlete, having run in 120 marathons in cities around the world including New York, Berlin, Boston, Chicago, London, Paris, Washington, and San Diego. Michael was born in Bocholt, Germany in 1968, and educated in Germany, Spain, and the UK in business and finance. He has since worked for Volkswagen, General Motors, Mitsubishi Motors, Nikola Corporation, and several other players in the automotive industry, in various C-suite capacities.

Ulrike Louis—Michael's wife and co-author—was born in Schwerin, Germany. She is a practicing dentist, hypnotherapist, businesswoman, and trained nurse with an interest in nutrition, healthy living, and wellness. As a married couple, Michael and Ulrike share each other's passion for living in harmony with nature and ending climate change.

They have two children pursuing dentistry and data science respectively. The couple enjoy bike rides, experiencing the great outdoors, and spending time with their dog.

REFERENCES

[1] Forbes. "QuotesThoughts On The Business Of Life". Forbes. (n.d.). https://www.forbes.com/quotes/1164/

[2] 2. Pande, Shania. "Climate Change: Research Gaps and Policy Priorities". Meeting Report. Online Webinar. Physiological Society. 2022, June 15. https://www.physoc.org/magazine-articles/climate-change-research-gaps-and-policy-priorities/

[3] 3. The Nature Conservancy. "Calculate Your Carbon Footprint". Nature. (n.d.). https://www.nature.org/en-us/get-involved/how-to-help/carbon-footprint-calculator/

[4] National Grid. "What are greenhouse gasses?". Energy Explained. National Grid. 2023, February 23. https://www.nationalgrid.com/stories/energy-explained/what-are-greenhouse-gases

[5] . NASA. "Climate Forcings and Global Warming". Earth Observatory. NASA. 2009, January 14. https://Earthobservatory.nasa.gov/features/EnergyBalance/page7.php

[6] National Grid. "What are greenhouse gasses?". Energy Explained. National Grid. 2023, February 23. https://www.nationalgrid.com/stories/energy-explained/what-are-greenhouse-gases

[7] U.S. Global Change Research Program. "Atmospheric Carbon Dioxide". Indicators. U.S. Global Change Research Program. (n.d.). https://www.globalchange.gov/indicators/atmospheric-carbon-dioxide#:~:text=Carbon%20Dioxide%20in%20the%20Atmosphere,the%20industrial%20revolution%20(1750)

[8] Santer, Benjamin D. Painter, Jeffrey F. Mears, Carl A. Doutriaux Caldwell, Peter. Arblaster, Julie M. Cameron-Smith, P.J. Gillet, Nathan P. Glecker Peter J. Lanzante, John. Perlwitz, Judith. Solomon, Susan. Stott, Peter A. Taylor, Karl E. Terray, Laurent. Thorne, Peter W. Wehner, Michael F. Wentz, Frank J. Wigley, Tom M. L. Wilcox, Laura. Zou, Cheng-Zhi. "Identifying human influences on atmospheric temperature". Computer Science. Lawrence Berkeley National Laboratory. eScholarship.org. 2014. https://escholarship.org/uc/item/203472j8

[9] Santer, Benjamin D. Painter, Jeffrey F. Mears, Carl A. Doutriaux Caldwell, Peter. Arblaster, Julie M. Cameron-Smith, P.J. Gillet, Nathan P. Glecker Peter J. Lanzante, John. Perlwitz, Judith. Solomon, Susan. Stott, Peter A. Taylor, Karl E. Terray, Laurent. Thorne, Peter W. Wehner, Michael F. Wentz, Frank J. Wigley, Tom M. L. Wilcox, Laura. Zou, Cheng-Zhi. ""Identifying human

influences on atmospheric temperature". Computer Science. Lawrence Berkeley National Laboratory. eScholarship.org. 2014. https://escholarship.org/uc/item/203472j8

10 NASA. "How is Today's Warming Different from the Past?". Global Warming. NASA. 2010, June 3. https://Earthobservatory.nasa.gov/features/GlobalWarming/page3.php

[11] International Energy Agency. CO2 Emissions in 2022". International Energy Agency. 2023, March. https://iea.blob.core.windows.net/assets/3c8fa115-35c4-4474-b237-1b00424c8844/CO2Emissionsin2022.pdf

[12] International Energy Agency. CO2 Emissions in 2022". International Energy Agency. 2023, March. https://iea.blob.core.windows.net/assets/3c8fa115-35c4-4474-b237-1b00424c8844/CO2Emissionsin2022.pdf

[13] London School of Economics and Political Science. "What is the role of deforestation in climate change and how can 'Reducing Emissions from Deforestation and Degradation' (REDD+) help?". Grantham Research Institute on Climate Change and the Environment. London School of Economics. 2023, February 10. https://www.lse.ac.uk/granthaminstitute/explainers/whats-redd-and-will-it-help-tackle-climate-change/

[14] Britannica. "Photosynthesis". Britannica. 2023, October 15. Britannica. https://www.britannica.com/science/photosynthesis

[15] NASA. "How Plants Can Change Our Climate". Earth Observatory. NASA. 2002, May 6. https://Earthobservatory.nasa.gov/features/LAI/LAI2.php#:~:text=As%20plants%20'breathe'%20and%20',in%20the%20process%20of%20photosynthesis.

[16] Directorate-General for Environment. "Field to fork: global food miles generate nearly 20% of all CO2 emissions from food". Europa. 2023, January 25. https://environment.ec.europa.eu/news/field-fork-global-food-miles-generate-nearly-20-all-co2-emissions-food-2023-01-25_en#:~:text=The%20researchers%20also%20estimated%20the,of%20the%20world's%20GHG%20emissions.

[17] Food and Agriculture Organization of the United Nations. "Emissions due to agriculture. Global, regional and country trends 1990–2018". FAOSTAT Analytical Brief 18. Food and Agriculture Organization of the United Nations. 2021. https://www.fao.org/3/cb3808en/cb3808en.pdf

[18] Food and Agriculture Organization of the United Nations. "Emissions due to agriculture. Global, regional and country trends 1990–2018". FAOSTAT

Analytical Brief 18. Food and Agriculture Organization of the United Nations. 2021. https://www.fao.org/3/cb3808en/cb3808en.pdf

[19] Volcano Hazards Program. "Carbon dioxide gas can collect in low-lying volcanic areas, posing ...". Images. USGS. 1989, February 14. https://www.usgs.gov/programs/VHP/volcanic-gases-can-be-harmful-health-vegetation-and-infrastructure

[20] Volcano Hazards Program. "Carbon dioxide gas can collect in low-lying volcanic areas, posing ...". Images. USGS. 1989, February 14. https://www.usgs.gov/programs/VHP/volcanic-gases-can-be-harmful-health-vegetation-and-infrastructure

[21] 21. Lindsey, Rebecca. Dahlman, Luann. "Climate Change: Global Temperature". Climate.gov. 2023, January 18. https://www.climate.gov/news-features/understanding-climate/climate-change-global-temperature

[22] 22. Lindsey, Rebecca. Dahlman, Luann. "Climate Change: Global Temperature". Climate.gov. 2023, January 18. https://www.climate.gov/news-features/understanding-climate/climate-change-global-temperature

[23] 23. Stephan. "What ocean heating reveals about global warming". Climate Modelling. Real Climate. 2013, September 25. https://www.realclimate.org/index.php/archives/2013/09/what-ocean-heating-reveals-about-global-warming/

[24] 24. Stephan. "What ocean heating reveals about global warming". Climate Modelling. Real Climate. 2013, September 25. https://www.realclimate.org/index.php/archives/2013/09/what-ocean-heating-reveals-about-global-warming/

[25] 25. IPCC. "Climate Change 2007: Working Group I: The Physical Science Basis". Fourth Assessment Report: Climate Change 2007. IPCC. 2007. https://archive.ipcc.ch/publications_and_data/ar4/wg1/en/faq-5-1.html

[26] 26. IPCC. "Climate Change 2007: Working Group I: The Physical Science Basis". Fourth Assessment Report: Climate Change 2007. IPCC. 2007. https://archive.ipcc.ch/publications_and_data/ar4/wg1/en/faq-5-1.html

[27] 27. Woods Hole Oceanographic Institution. "Ocean Acidification". Ocean Topics. Woods Hole Oceanographic Institution. (n.d.). https://www.whoi.edu/know-your-ocean/ocean-topics/how-the-ocean-works/ocean-chemistry/ocean-acidification/

[28] 28. UN Climate Change. "The Paris Agreement". Process and Meetings. UNFCCC. (n.d.). https://unfccc.int/process-and-meetings/the-paris-agreement

[29] 29. UN Climate Change. "The Paris Agreement". Process and Meetings. UNFCCC. (n.d.). https://unfccc.int/process-and-meetings/the-paris-agreement

30 UN Climate Change. "The Paris Agreement". Process and Meetings. UNFCCC. (n.d.). https://unfccc.int/process-and-meetings/the-paris-agreement

31 Victor, David G. Lumkowsky, Marcel. Dannenberg, Astrid. Carlton, Emily. "Success of the Paris Agreement hinges on the credibility of national climate goals". Articles. Brookings. 2022, September 30. https://www.brookings.edu/articles/success-of-the-paris-agreement-hinges-on-the-credibility-of-national-climate-goals/

32 Maizland, Lindsay. "Global Climate Agreements: Successes and Failures". Backgrounder. Council on Foreign Relations. 2023, September 15. https://www.cfr.org/backgrounder/paris-global-climate-change-agreements

33 Milman, Oliver. "The Governments Falling Woefully Short of Paris Climate Pledges, Study Finds". The Guardian. 2021, September 15. https://www.theguardian.com/science/2021/sep/15/governments-falling-short-paris-climate-pledges-study#:~:text=Climate%20pledges%20made%20by%20Russia,those%20deemed%20%E2%80%9Chighly%20insufficient%E2%80%9D

34 34. World Wildlife Fund. "What's The Difference Between Climate Change Mitigation and Adaptation?". Stories. WWF. (n.d.). https://www.worldwildlife.org/stories/what-s-the-difference-between-climate-change-mitigation-and-adaptation

35 World Wildlife Fund. "What's The Difference Between Climate Change Mitigation and Adaptation?". Stories. WWF. (n.d.). https://www.worldwildlife.org/stories/what-s-the-difference-between-climate-change-mitigation-and-adaptation

36 GoodReads. "..Yes, but I have something he will never have … enough..". Quotable Quotes. GoodReads. (n.d.). .https://www.goodreads.com/quotes/10651136-at-a-party-given-by-a-billionaire-on-shelter-island

37 Cozzi, Laura. Chen, Olivia. Kim, Hyeji. "The world's top 1% of emitters produce over 1000 times more CO2 than the bottom 1%". Commentaries. International Energy Agency. 2023, February 22. https://www.iea.org/commentaries/the-world-s-top-1-of-emitters-produce-over-1000-times-more-co2-than-the-bottom-1

38 UCAR - Center for Science Education. "What's Your Carbon Footprint?".(n.d.). https://scied.ucar.edu/learning-zone/climate-solutions/carbon-footprint#:~:text=Worldwide%2C%20the%20average%20person%20produces,causes%20our%20climate%20to%20warm.

39. AJLabs. "How much does Africa contribute to global carbon emissions?". Explainer Al Jazeera. 2023, September 4. https://www.aljazeera.com/news/2023/9/4/how-much-does-africa-

contribute-to-global-carbon-emissions#:~:text=On%20a%20per%20capita%20basis,North%20America%20(10.3%20tonnes).

[40] Cozzi, Laura. Chen, Olivia. Kim, Hyeji. "The world's top 1% of emitters produce over 1000 times more CO2 than the bottom 1%". Commentaries. International Energy Agency. 2023, February 22. https://www.iea.org/commentaries/the-world-s-top-1-of-emitters-produce-over-1000-times-more-co2-than-the-bottom-1

[41] Rescue.org. "10 countries at risk of climate disaster". Climate Change. Rescue. 2023, November 3. https://www.rescue.org/article/10-countries-risk-climate-disaster

[42] Rescue.org. "10 countries at risk of climate disaster". Climate Change. Rescue. 2023, November 3. https://www.rescue.org/article/10-countries-risk-climate-disaster

[43] United Nations - Environment Programme. "Tracking Progress". Global Alliance for Buildings and Construction. (n.d.). https://globalabc.org/our-work/tracking-progress-global-status-report

[44] United Nations - Environment Programme. "Tracking Progress". Global Alliance for Buildings and Construction. (n.d.). https://globalabc.org/our-work/tracking-progress-global-status-report

[45] Dark Matter Labs. "Designing Our Future". A European Bauhaus Economy. Desire. Irresistible Circular Society. Dark Matter Labs. 2023, July 7. https://www.irresistiblecircularsociety.eu/assets/uploads/20230707-New-European-Bauhaus-Economy_Digital-version_DML.pdf

[46] Dark Matter Labs. "Designing Our Future". A European Bauhaus Economy. Desire. Irresistible Circular Society. Dark Matter Labs. 2023, July 7. https://www.irresistiblecircularsociety.eu/assets/uploads/20230707-New-European-Bauhaus-Economy_Digital-version_DML.pdf

[47] Dark Matter Labs. "Designing Our Future". A European Bauhaus Economy". Desire. Irresistible Circular Society. Dark Matter Labs. 2023, July 7. https://www.irresistiblecircularsociety.eu/assets/uploads/20230707-New-European-Bauhaus-Economy_Digital-version_DML.pdf

[48] Gero Reuter. "Climate-friendly, affordable housing: Is it possible?". Global Issues. DW. 2023, June 1. https://www.dw.com/en/climate-friendly-affordable-housing-is-it-possible/a-64245802

[49] United Nations Environment Programme. "2022 Global Status Report for Buildings and Construction". Publication. United Nations Environment Programme. 2022, November 9. https://www.unep.org/resources/publication/2022-global-status-report-buildings-and-construction

[50] Study.com. "Types of Houses | Overview & Examples". Chapter - Study.com. . (n.d.). https://study.com/academy/lesson/types-of-housing-overview-examples.html

[51] Zinn, Dori. "What Is A Single-Family Home?". Mortgages. Forbes. 2022, October 14 (Updated). https://www.forbes.com/advisor/mortgages/real-estate/single-family-home/

[52] The Britannica Dictionary. "Apartment". Britannica. (n.d.). https://www.britannica.com/dictionary/apartment

[53] Ullrich, Jess. "What is an apartment?". Bank Rate. 2022, October 18. https://www.bankrate.com/real-estate/what-is-an-apartment/

[54] Study. "Condominium | Definition, Advantages & Disadvantages". Study.com. (n.d.). https://study.com/learn/lesson/what-is-a-condominium.html

[55] Bond, Casey. Fontinelle, Amy. "What is a Manufactured Home". Mortgages. Forbes. https://www.forbes.com/advisor/mortgages/manufactured-homes/

[56] Smith, Lisa. "Housing Cooperatives: A Unique Type of Home Ownership". Real Estate Investing. Investopedia. 2022, January 14. https://www.investopedia.com/articles/pf/08/housingco-op.asp

[57] The Carbon Leadership Forum. "Life Cycle Assessment of Buildings: A Practical Guide". Carbon Leadership Forum. 2019, June. https://carbonleadershipforum.org/wp-content/uploads/2019/05/CLF-LCA-Practice-Guide_2019-05-23.pdf

[58] United Nations. "Population".Global Issues. United Nations. (n.d.). https://www.un.org/en/global-issues/population#:~:text=Our%20growing%20population&text=The%20world's%20population%20is%20expected,billion%20in%20the%20mid%2D2080s.

[59] The Conversation. "The world needs to build more than two billion new homes over the next 80 years". The Conversation. 2018, February 28. https://theconversation.com/the-world-needs-to-build-more-than-two-billion-new-homes-over-the-next-80-years-91794

[60] The Conversation. "The world needs to build more than two billion new homes over the next 80 years". The Conversation. 2018, February 28. https://theconversation.com/the-world-needs-to-build-more-than-two-billion-new-homes-over-the-next-80-years-91794

[61] The Conversation. "The world needs to build more than two billion new homes over the next 80 years". The Conversation. 2018, February 28. https://theconversation.com/the-world-needs-to-build-more-than-two-billion-new-homes-over-the-next-80-years-91794

[62] Badger, Emily. "Why We Should be Worried About the Rapid Growth in Global Households". Housing. City Lab. Bloomberg. 2014, February 14.

https://www.bloomberg.com/news/articles/2014-02-14/why-we-should-be-worried-about-the-rapid-growth-in-global-households

[63] The Conversation. "The world needs to build more than two billion new homes over the next 80 years". The Conversation. 2018, February 28. https://theconversation.com/the-world-needs-to-build-more-than-two-billion-new-homes-over-the-next-80-years-91794

[64] The Conversation. "The world needs to build more than two billion new homes over the next 80 years". The Conversation. 2018, February 28. https://theconversation.com/the-world-needs-to-build-more-than-two-billion-new-homes-over-the-next-80-years-91794

[65] The Conversation. "The world needs to build more than two billion new homes over the next 80 years". The Conversation. 2018, February 28. https://theconversation.com/the-world-needs-to-build-more-than-two-billion-new-homes-over-the-next-80-years-91794

[66] Badger, Emily. "Why We Should be Worried About the Rapid Growth in Global Households". Housing. City Lab. Bloomberg. 2014, February 14. https://www.bloomberg.com/news/articles/2014-02-14/why-we-should-be-worried-about-the-rapid-growth-in-global-households

[67] Hendricks, Scotty. "Study: Big homes have big carbon footprints". Big Think. 2020, July 27. https://bigthink.com/the-present/housing-greenhouse-gases/

[68] Cho, Renee. "Heating Buildings Leaves a Huge Carbon Footprint, But There's a Fix For It". Climate. Columbia University. 2019, January 15. https://news.climate.columbia.edu/2019/01/15/heat-pumps-home-heating/

[69] Mollenkamp, Daniel Thomas. "What is Sustainability? How Sustainabilities Work, Benefits, and Example". Terms. Investopedia. 2023, December 13. https://www.investopedia.com/terms/s/sustainability.asp

[70] UN-Habitat. "The Climate is Changing, So must our Homes & How We Build Them". News. UN Habitat. (n.d.). https://unhabitat.org/news/23-sep-2019/the-climate-is-changing-so-must-our-homes-how-we-build-them

[71] UN-Habitat. "The Climate is Changing, So must our Homes & How We Build Them". News. UN Habitat. (n.d.). https://unhabitat.org/news/23-sep-2019/the-climate-is-changing-so-must-our-homes-how-we-build-them

[72] UN-Habitat World. "SHERPA For Sustainable Housing Projects". UN-Habitat World. YouTube. 2018, May 31. https://www.youtube.com/watch?app=desktop&v=DyqMBL5ymX0

[73] Active Sustainability. "Non-Polluting Construction Materials". Active Sustainability. (n.d.). https://www.activesustainability.com/construction-and-urban-development/non-polluting-construction-materials/?_adin=02021864894

[74] Active Sustainability. "Non-Polluting Construction Materials". Active Sustainability. (n.d.). https://www.activesustainability.com/construction-and-urban-development/non-polluting-construction-materials/?_adin=02021864894

[75] Mitra, B.. (2014). Environment-Friendly Composite Materials: Biocomposites and Green Composites. Defence Science Journal. 2014. https://core.ac.uk/download/pdf/333722313.pdf.

[76] S. Sair, B. Mandili, M. Taqi, A. El Bouari. "Development of a new eco-friendly composite material based on gypsum reinforced with a mixture of cork fibre and cardboard waste for building thermal insulation, Composites Communications". Volume 16, 2019. Pages 20-24. ISSN 2452-2139. https://doi.org/10.1016/j.coco.2019.08.010.

[77] Construction & Civil Engineering Magazine. "10 Sustainable Building Materials". Construction & Civil Engineering Magazine. 2023, June 12. https://ccemagazine.com/news/10-sustainable-building-materials/

[78] Britannica. "Bamboo". Arts & Culture. Britannica. 2023, November 9. https://www.britannica.com/plant/bamboo

[79] Hempcrete Cymru. "What is Hempcrete?". Hempcrete Cymru.(n.d.). https://www.hempcretecymru.com/what-is-hempcrete

[80] Nortsar. "What is Steel Recycling?". Norstar. (n.d.). https://www.norstar.com.au/what-is-steel-recycling/

[81] Britannica. "Rammed Earth". Science & Tech. Britannica. (n.d.). https://www.britannica.com/technology/rammed-Earth

[82] UN Climate Technology Centre & Network. "Glass recyclingg". UN Environment Programme. UN Climate Technology Centre & Network. (n.d.). https://www.ctc-n.org/technologies/glass-recycling

[83] Kubba Ph.D. "Chapter 6 - Green Materials and Products". LEED Practices, Certification, and Accreditation Handbook". Science Direct. 2010. https://www.sciencedirect.com/science/article/abs/pii/B9781856176910000060

[84] Kubba Ph.D. "Chapter 6 - Green Materials and Products". LEED Practices, Certification, and Accreditation Handbook". Science Direct. 2010. https://www.sciencedirect.com/science/article/abs/pii/B9781856176910000060

[85] Anthology Woods. "The Difference Between Reclaimed and Salvaged Wood". Anthology Woods. 2022, October 6. https://anthologywoods.com/aw-blog/the-difference-between-reclaimed-and-salvaged-wood

[86] Evan. "What Building Material (Wood, Steel, Concrete), Has The Smallest Overall Environment Impact?". Debating Science. University of Massachusetts

- Amherst. 2016, April 21. https://blogs.umass.edu/natsci397a-eross/what-building-material-wood-steel-concrete-has-the-smallest-overall-environment-impact/

[87] Serhant, Ryan. "8 Eco-Friendly Features To Add To Your Home". Forbes. 2021, November 9.
https://www.forbes.com/sites/ryanserhant/2021/11/09/8-eco-friendly-features-to-add-to-your-home/?sh=1fef4bdb591d

[88] Serhant, Ryan. "8 Eco-Friendly Features To Add To Your Home". Forbes. 2021, November 9.
https://www.forbes.com/sites/ryanserhant/2021/11/09/8-eco-friendly-features-to-add-to-your-home/?sh=1fef4bdb591d

[89] Serhant, Ryan. "8 Eco-Friendly Features To Add To Your Home". Forbes. 2021, November 9.
https://www.forbes.com/sites/ryanserhant/2021/11/09/8-eco-friendly-features-to-add-to-your-home/?sh=1fef4bdb591d

[90] Permaculture. "What is Permaculture?". Permaculture. (n.d.). https://www.permaculture.co.uk/what-is-permaculture/

[91] Serhant, Ryan. "8 Eco-Friendly Features To Add To Your Home". Forbes. 2021, November 9.
https://www.forbes.com/sites/ryanserhant/2021/11/09/8-eco-friendly-features-to-add-to-your-home/?sh=1fef4bdb591d

[92] Morrison, Rose. "Are Here: Why Aren't They More Popular". Earth.org. 2023, April 26. https://Earth.org/sustainable-housing/

[93] Collins. "Prefab". Collins Dictionary. (n.d.). https://www.collinsdictionary.com/dictionary/english/prefab

[94] Merriam-Webster. "Tiny House". Merriam-Webster. (n.d.). https://www.merriam-webster.com/dictionary/tiny%20house

[95] Rippon, Jordan A. "The Benefits and Limitations of Prefabricated Home Manufacturing in North America". University of British Columbia. 2011, April 11. https://open.library.ubc.ca/media/stream/pdf/52966/1.0103127/3

[96] Terezakis, Peter. "Earth Ships". Green World. New York University. 2022, November 14. https://wp.nyu.edu/gw/Earth-ships/

[97] Serhant, Ryan. "8 Eco-Friendly Features To Add To Your Home". Forbes. 2021, November 9.
https://www.forbes.com/sites/ryanserhant/2021/11/09/8-eco-friendly-features-to-add-to-your-home/?sh=1fef4bdb591d

[98] Serhant, Ryan. "8 Eco-Friendly Features To Add To Your Home". Forbes. 2021, November 9.
https://www.forbes.com/sites/ryanserhant/2021/11/09/8-eco-friendly-features-to-add-to-your-home/?sh=1fef4bdb591d

[99] Rogers, Heather. "Current Thinking". Reconsideration. The New York Times Magazine. 2007, June 3. https://www.nytimes.com/2007/06/03/magazine/03wwln-essay-t.html

[100] World Data.info. "Sunrise and sunset in the Netherlands". WorldData.info. (n.d.). https://www.worlddata.info/europe/netherlands/sunset.phpp

[101] Dutch News. "Vandebron to charge customers fees for surplus solar power". Dutch News. 2023, August 15. https://www.dutchnews.nl/2023/08/vandebron-to-charge-customers-fees-for-surplus-solar-power/

[102]. Vora, Shivani. "5 Properties Bringing New Life to Biophilic Design". Architecture + Design. Architectural Design. Architectural Digest. 2022, January 8. https://www.architecturaldigest.com/story/biophilic-design-movement

[103]. Kellert ,Stephen R. Calabrese, Elizabeth F. "The Practice of Biophilic Design". University of Minnesota. 2015. https://biophilicdesign.umn.edu/sites/biophilic-net-positive.umn.edu/files/2021-09/2015_Kellert%20_The_Practice_of_Biophilic_Design.pdf

[104]. New Zealand. "About The Hobbit Trilogy". New Zealand. (n.d.). https://www.newzealand.com/int/feature/about-the-hobbit-trilogy/

[105]. McLennan, Leah. "When it comes to hobbits, Tolkien fans would do well to visit Iceland". Adventure. News.com.au. 2015, October 21. https://www.news.com.au/travel/travel-ideas/adventure/when-it-comes-to-hobbits-tolkien-fans-would-do-well-to-visit-iceland/news-story/76de4572652e636c34d0023666f512e7

[106]. McLennan, Leah. "When It Comes to Hobbits, Tolkien Fans Would Do Well to Visit Iceland". News.com.au. 2015, October 21. https://www.news.com.au/travel/travel-ideas/adventure/when-it-comes-to-hobbits-tolkien-fans-would-do-well-to-visit-iceland/news-story/76de4572652e636c34d0023666f512e7

[107]. Energy. "Efficient Earth Sheltered Homes". Energy. (n.d.). https://www.energy.gov/energysaver/efficient-Earth-sheltered-homes#:~:text=Such%20a%20house%20is%20built,stairway%20from%20the%20ground%20level.

[108]. Energy. "Efficient Earth Sheltered Homes". Energy. (n.d.). https://www.energy.gov/energysaver/efficient-Earth-sheltered-homes#:~:text=Such%20a%20house%20is%20built,stairway%20from%20the%20ground%20level.

[109]. Energy. "Efficient Earth Sheltered Homes". Energy. (n.d.). https://www.energy.gov/energysaver/efficient-Earth-sheltered-homes#:~:text=Such%20a%20house%20is%20built,stairway%20from%20the%20ground%20level.

110. Cambridge Dictionary. "Passive House". Cambridge Dictionary. (n.d.). https://dictionary.cambridge.org/dictionary/english/passivhaus

111. The Investopedia Team. "Sharing Economy: Model Defined, Criticisms, and How It's Evolving". Sectors & Industries. Investopedia. 2023, December 17. https://www.investopedia.com/terms/s/sharing-economy.asp

112. Marshall McCANN Architects. "Why Build a Passive House? Passive House Benefits". Marshall McCANN Architects. (n.d.).. https://mmccarchitects.com/why-build-a-passive-house-passive-house-benefits/

113. Tobias, Michael. "Passive House Design & Construction: Pros & Cons". Blog. NY Engineers. https://www.ny-engineers.com/blog/green-building-trends-pros-cons-of-passive-house-construction

114. Cambridge Dictionary. "Sharing". Cambridge. (n.d.). https://dictionary.cambridge.org/dictionary/english/sharing

115. The Investopedia Team. "Sharing Economy: Model Defined, Criticisms, and How It's Evolving". Sectors & Industries. Investopedia. 2023, December 17. https://www.investopedia.com/terms/s/sharing-economy.asp

116. The Investopedia Team. "Sharing Economy: Model Defined, Criticisms, and How It's Evolving". Sectors & Industries. Investopedia. 2023, December 17. https://www.investopedia.com/terms/s/sharing-economy.asp

117. Yaraghi, Niam. Ravi, Shamika. "The Current and Future State of the Sharing Economy". Brookings. 2016, December 29. https://www.brookings.edu/articles/the-current-and-future-state-of-the-sharing-economy/

118. Chen, Yujia. Schuckert, Markus. "Why People Choose Airbnb Over Hotel?". Conference: The 80th TOSOK Gangwon Pyeongchang International Tourism ConferenceAt: Pyeongchang, Gangwon Province. The Hong Kong Polytechnic University. 2016, July. https://www.researchgate.net/publication/305225839_Why_people_choose_Airbnb_over_Hotel

119. Arizona State University. "The Gentle Science of Persuasion, Part Three: Social Proof". Research. Arizona State University. (n.d.). https://news.wpcarey.asu.edu/20070103-gentle-science-persuasion-part-three-social-proof

120. Host a Sister. Host a Sister - Public Page. Facebook.com. (n.d.). https://web.facebook.com/hostasister

121. CouchSurfing. "Home page". CouchSurfing. (n.d.). https://www.couchsurfing.com/

122. HostelWorld. "Understanding the Carbon Impact of Hostels vs Hotel". BureaVeritaz.co.uk. 2022.

https://www.bureauveritas.co.uk/sites/g/files/zypfnx216/files/media/document/Hostel%20World%20FINAL%20REPORT%20%281%29.pdf

[123]. GoodReads. "Quotable Quote: "Wherever you go, go with all your heart.—Confucius".GoodReads. (n.d.). https://www.goodreads.com/quotes/8496573-wherever-you-go-go-with-all-your-heart

[124]. GoodReads. "Recycling Quotes: [T]his readiness to assume the guilt for the threats to our environment is deceptively reassuring…". Good Reads. (n.d.). https://www.goodreads.com/quotes/tag/recycling

[125]. WorldoMeter. "Current World Population. World Meter. (n.d.). https://www.worldometers.info/world-population

[126]. Our World in Data. "Fertility Rate". Our World in Data. (n.d.). https://ourworldindata.org/fertility-rate

[127]. Our World in Data. "Fertility Rate". Our World in Data. (n.d.). https://ourworldindata.org/fertility-rate

[128]. GoodReads. "Past Quotes: "Study the past if you would define the future." — Confucius. GoodReads. (n.d.). https://www.goodreads.com/quotes/tag/past

[129]. GoodReads. "Quotable Quotes: "The Only Thing That Is Constant Is Change -" — Heraclitus". Good Reads. (n.d.). https://www.goodreads.com/quotes/336994-the-only-thing-that-is-constant-is-change--

[130]. Heliyon. "Environmental effects of COVID-19 pandemic and potential strategies of sustainability". PubMed Central. 2020 September 17. https://www.ncbi.nlm.nih.gov/pmc/articles/PMC7498239/#:~:text=The%20global%20disruption%20caused%20by,different%20parts%20of%20the%20world.

[131]. Ronaghi, Marzieh. Scorsone, Eric. "The Impact of COVID-19 Outbreak on CO2 Emissions in the Ten Countries with the Highest Carbon Dioxide Emissions". Articles. J Environ Public Health. https://www.ncbi.nlm.nih.gov/pmc/articles/PMC10281825/#:~:text=The%20dramatic%20reduction%20in%20carbon,atmospheric%20carbon%20dioxide%20is%20observed.

[132]. Ronaghi, Marzieh. Scorsone, Eric. "The Impact of COVID-19 Outbreak on CO2 Emissions in the Ten Countries with the Highest Carbon Dioxide Emissions". Articles. J Environ Public Health. https://www.ncbi.nlm.nih.gov/pmc/articles/PMC10281825/#:~:text=The%20dramatic%20reduction%20in%20carbon,atmospheric%20carbon%20dioxide%20is%20observed.

[133]. Montt, Guillermo. Fraga, Federico. Harsdorff. "The Future of Work in a Changing Natural Environment: Climate Change, degradation and Sustainability". International Labour Organization. 2018.

https://www.ilo.org/wcmsp5/groups/public/---dgreports/---cabinet/documents/publication/wcms_644145.pdf

[134] Montt, Guillermo. Fraga, Federico. Harsdorff, Marek. "The future of work in a changing natural environment: Climate change, degradation and sustainability". International Labour Organization. 2018. https://www.ilo.org/wcmsp5/groups/public/---dgreports/---cabinet/documents/publication/wcms_644145.pdf

[135]. World Food Programme. "Global report on food crises: Number of people facing acute food insecurity rose to 258 million in 58 countries in 2022". News Release. World Food Programme. 2023, May 3. https://www.wfp.org/news/global-report-food-crises-number-people-facing-acute-food-insecurity-rose-258-million-58.

[136]. World Food Programme. "Global report on food crises: Number of people facing acute food insecurity rose to 258 million in 58 countries in 2022". News Release. World Food Programme. 2023, May 3. https://www.wfp.org/news/global-report-food-crises-number-people-facing-acute-food-insecurity-rose-258-million-58.

[137]. United Nations. "Causes and Effects of Climate Change". Climate Action. United Nations. (n.d.). https://www.un.org/en/climatechange/science/causes-effects-climate-change#:~:text=Most%20cars%2C%20trucks%2C%20ships%2C,gasoline%2C%20in%20internal%20combustion%20engines.

[138]. United Nations. "Causes and Effects of Climate Change". Climate Action. United Nations. (n.d.). https://www.un.org/en/climatechange/science/causes-effects-climate-change

[139]. University of Gothenburg. "Machines and Lighting Affect the Climate". The Faculty of Science. University of Gothenburg. 2020, January 23. https://www.gu.se/en/news/machines-and-lighting-affect-the-climate#:~:text=Impact%20of%20machines%20and%20equipment,a%20global%20scale%20so%20far.

[140]. University of Gothenburg. "Machines and Lighting Affect the Climate". The Faculty of Science. University of Gothenburg. 2020, January 23. https://www.gu.se/en/news/machines-and-lighting-affect-the-climate#:~:text=Impact%20of%20machines%20and%20equipment,a%20global%20scale%20so%20far.

[141]. University of Gothenburg. "Machines and Lighting Affect the Climate". The Faculty of Science. University of Gothenburg. 2020, January 23. https://www.gu.se/en/news/machines-and-lighting-affect-the-climate#:~:text=Impact%20of%20machines%20and%20equipment,a%20global%20scale%20so%20far.

142. Cox, Hugo. "Data Centres: Smart Solution or Eco Nightmare?". Modus. 2020, July 30. https://ww3.rics.org/uk/en/modus/technology-and-data/harnessing-data/how-the-cloud-created-the-perfect-storm.html

143. Cox, Hugo. "Data Centres: Smart Solution or Eco Nightmare?". Modus. 2020, July 30. https://ww3.rics.org/uk/en/modus/technology-and-data/harnessing-data/how-the-cloud-created-the-perfect-storm.html

144. National Geographic. "Non-renewable energy". Article. National Geographic. (n.d.). https://education.nationalgeographic.org/resource/non-renewable-energy/

145. UC Davis. "What is Solar Power". Definitions. UC Davis. 2021, November 5. https://www.ucdavis.edu/climate/definitions/how-is-solar-power-generated

146. College of Earth and Mineral Sciences. "EGEE 102: Energy Conservation for Environmental Protection". "Sources of Energy". Pennsylvania State. (n.d.). https://www.e-education.psu.edu/egee102/node/1909

147. US Energy Information Administration.. "Biomass - Renewable Energy from Plants and Animals". Basics. US Energy Information Administration. 2023, June 20. https://www.eia.gov/energyexplained/biomass/

148. Clean Energy Ideas. "8 Examples of Biomass". Biomass. Clean Energy Ideas. 2019, May 27. https://www.clean-energy-ideas.com/biomass/bioenergy/8-examples-of-biomass/

149. US Energy Information Administration. "Biomass Explained: Biomass and the Environment". Basics. US Energy Information Administration. 2022, November 7. https://www.eia.gov/energyexplained/biomass/biomass-and-the-environment.php#:~:text=Burning%20either%20fossil%20fuels%20or,a%20carbon%2Dneutral%20energy%20source.

150. Dinan, Georgeanne. "Treating biomass as a carbon-neutral energy source will only drive further wood harvests and carbon emissions at a time when reforestation and decarbonization are critical.". Renewables, Climate. Kleinman Center for Energy Policy. University of Pennsylvania. 2021, June 18. https://kleinmanenergy.upenn.edu/news-insights/biomass-energy-climate-solution-or-potential-catastrophe/

151. UC Davis. "What is Solar Power". Definitions. UC Davis. (n.d.). https://www.ucdavis.edu/climate/definitions/how-is-solar-power-generated

152. UC Davis. "What is Solar Power". Definitions. UC Davis. (n.d.). https://www.ucdavis.edu/climate/definitions/how-is-solar-power-generated

153. International Energy Agency. "Hydroelectricity". Renewables. International Energy Agency. (n.d.). https://www.iea.org/energy-system/renewables/hydroelectricity

154. Office of Energy Efficiency & Renewable Energy. "Hydropower Basics". Water Power Technologies Office. Office of Energy Efficiency & Renewable Energy. (n.d.). https://www.energy.gov/eere/water/hydropower-basics#:~:text=What%20is%20Hydropower%3F,moving%20water%20to%20generate%20electricity.

155. Manawka Tefera, Wasu. Kasiviswanathan, K.S. "A Global-Scale Hydropower Potential Assessment and Feasibility Evaluations". Water Resources and Economics. Elsevier. Volume 38. Science Direct. 2022, April. https://www.sciencedirect.com/science/article/abs/pii/S2212428422000068#:~:text=Globally%20a%20theoretical%20hydropower%20potential,30%25%20flow%20dependability%20was%20estimated.

156. International Energy Agency. "Hydroelectricity". Renewables. International Energy Agency. (n.d.). https://www.iea.org/energy-system/renewables/hydroelectricity

157 Office of Energy Efficiency & Renewable Energy. "Hydropower Basics". Water. Office of Energy Efficiency & Renewable Energy. (n.d.). https://www.energy.gov/eere/water/hydropower-basics

158. Pacific NorthWest National Laboratory. "Tidal Energy". Explainer Articles. Pacific NorthWest National Laboratory. (n.d.). https://www.pnnl.gov/explainer-articles/tidal-energy#:~:text=Tidal%20energy%20is%20a%20form,the%20water%20to%20move%20faster.

159 Boucher, Lauren. "What are the Pros and Cons of Hydropower and Tidal Energy". PopEd Blog. Population Education. 2015, July 13. https://populationeducation.org/what-are-pros-and-cons-hydropower-and-tidal-energy

160 Polaris Market Research. "Wave and Tidal Energy Market Share, Size, Trends, Industry Analysis Report, By Type (Wave Energy and Tidal Energy); By Application; By Technology; By Region; Segment Forecast, 2022-2030". Wave and Tidal Energy Market. Polaris Market Research. 2022, November. https://www.polarismarketresearch.com/industry-analysis/wave-and-tidal-energy-market#:~:text=For%20instance%2C%20as%20per%20the,world's%20total%20hydro%2Dpower%20capacity.

161. National Geographic. "Earth's core is the very hot, very dense center of our planet.". Core. Article. National Geographic. (n.d.). https://education.nationalgeographic.org/resource/core/

162. National Geographic. "Earth's core is the very hot, very dense center of our planet.". Core. Article. National Geographic. (n.d.). https://education.nationalgeographic.org/resource/core/

163. National Geographic. "Earth's core is the very hot, very dense center of our planet.". Core. Article. National Geographic. (n.d.). https://education.nationalgeographic.org/resource/core/

164. Energy Kid's Page. "Geothermal Energy". Energy. Lehigh University. 2008, July. https://ei.lehigh.edu/learners/energy/readings/geothermal.pdf

165. Robbins, Jim. "Can Geothermal Power Play a Key Role in the Energy Transition?". Yale Environment 360. Yale University. 2020, December 22. https://e360.yale.edu/features/can-geothermal-power-play-a-key-role-in-the-energy-transition

166. Robbins, Jim. "Can Geothermal Power Play a Key Role in the Energy Transition?". Yale Environment 360. Yale University. 2020, December 22. https://e360.yale.edu/features/can-geothermal-power-play-a-key-role-in-the-energy-transition

167. International Renewable Energy Agency. "Global Geothermal Market and Technology Assessment". International Renewable Energy Agency. 2023, February. https://www.irena.org/Publications/2023/Feb/Global-geothermal-market-and-technology-assessment

168. International Renewable Energy Agency. "Global Geothermal Market and Technology Assessment". Publications. International Renewable Energy Agency. 2023, February. https://www.irena.org/Publications/2023/Feb/Global-geothermal-market-and-technology-assessment

169. Robbins, Jim. Can Geothermal Power Play a Key Role in the Energy Transition?". Yale Environment 360. Yale University. 2022, December 22. https://e360.yale.edu/features/can-geothermal-power-play-a-key-role-in-the-energy-transition

170. Robbins, Jim. "Can Geothermal Power Play a Key Role in the Energy Transition?". Yale Environment 360. Yale University. 2020, December 22. https://e360.yale.edu/features/can-geothermal-power-play-a-key-role-in-the-energy-transition

171. Office of Energy Efficiency & Renewable Energy. "What is Wind Power?". Wind Energy Technology Office. Office of Energy Efficiency & Renewable Energy. (n.d.). https://windexchange.energy.gov/what-is-wind

172. Office of Energy Efficiency & Renewable Energy. "What is Wind Power?". Energy. Office of Energy Efficiency & Renewable Energy. (n.d.). https://windexchange.energy.gov/what-is-wind

173. US Energy Information Administration. "Wind Explained". Energy. (n.d.). https://www.eia.gov/energyexplained/wind/where-wind-power-is-harnessed.php

174. Statista. "Share of electricity generation from wind energy sources worldwide from 2010 to 2022". Energy Statista. Statista. 2023, July.

https://www.statista.com/statistics/1302053/global-wind-energy-share-electricity-mix/#:~:text=Wind%20energy%20sources%20accounted%20for,percent%20share%20a%20year%20earlier.

[175]. Penn State Extension. "What is Hydrogen Energy?". Articles. Pennsylvania State University. 2023, June 6. https://extension.psu.edu/what-is-hydrogen-energy

[176]. International Energy Agency. "Hydrogen". Overview. International Energy Agency. (n.d.). https://www.irena.org/Energy-Transition/Technology/Hydrogen

[177]. Energy Information Administration. "Hydrogen Explained". Energy. Energy Information Administration. (n.d.). https://www.eia.gov/energyexplained/hydrogen/use-of-hydrogen.php#:~:text=Hydrogen%20is%20currently%20used%20in,supplement%20or%20replace%20natural%20gas.

[178]. International Energy Administration. "Global Hydron Review 2023". Reports. Energy Information Administration. 2023. https://www.iea.org/reports/global-hydrogen-review-2023/executive-summary

[179]. International Energy Administration. "Hydrogen". Energy. Energy Information Administration. (n.d.). https://www.iea.org/energy-system/low-emission-fuels/hydrogen

[180]. Berners-Lee, Mike. "What's the carbon footprint of … building a house". Climate Crisis. The Guardian. 2010, October 14. https://www.theguardian.com/environment/green-living-blog/2010/oct/14/carbon-footprint-house

[181]. Evan. "What Building Material (Wood, Steel, Concrete) Has the Smallest Overall Environment Impact?". University of Massachusetts. 2016, April 21. https://blogs.umass.edu/natsci397a-eross/what-building-material-wood-steel-concrete-has-the-smallest-overall-environment-impact/

[182]. FAO. "Chapter 5: Construction Materials". Food and Agriculture Organization. (n.d.). https://www.fao.org/3/i2433e/i2433e02.pdf

[183]. Bialystok University of Technology. "3. Modern Building Materials". Bialystok University of Technology. {n.d.). https://pb.edu.pl/oficyna-wydawnicza/wp-content/uploads/sites/4/2018/12/Buildings-2020-part1-20.12-rozdz-3.pdf

[184]. Gagg, Colin R. "Cement and concrete as an engineering material: An historic appraisal and case study analysis". Engineering Failure Analysis. The Open University - United Kingdom. Elsevier. Science Direct. 2014, July 28. https://www.sciencedirect.com/science/article/abs/pii/S1350630714000387

[185]. University of Illinois - Urbana - Champaign. "Scientific Principles". Materials Science and Technology - Teacher's Workshop. University of Illinois Urbana-Champaign. (n.d.). http://matse1.matse.illinois.edu/concrete/prin.html

[186]. Arellano, Andrea. "Back to Basics: Cement vs Concrete". Giatec. 2022, November 30. https://www.giatecscientific.com/education/back-to-basics-cement-vs-concrete/#:~:text=In%20short%2C%20the%20difference%20between,water%2C%20sand%2C%20and%20rock.

[187]. Herff College of Engineering. "CIVL 1101 - Properties of Concrete". Civil Engineering. The University of Memphis. (n.d.). http://www.ce.memphis.edu/1101/notes/concrete/section_3_properties.html

[188]. Engineering for Change. "Cement: Enabler or Barrier for Global Development?". Articles. Engineering for Change. 2019, February 6. https://www.engineeringforchange.org/news/cement-enabler-barrier-global-development/

[189]. Engineering for Change. "Cement: Enabler or Barrier for Global Development?". Articles. Engineering for Change. 2019, February 6. https://www.engineeringforchange.org/news/cement-enabler-barrier-global-development/

[190]. Engineering for Change. "Cement: Enabler or Barrier for Global Development?". Articles. Engineering for Change. 2019, February 6. https://www.engineeringforchange.org/news/cement-enabler-barrier-global-development/

[191]. Food and Agriculture Organization. "Chapter 5: Construction Materials". Food and Agriculture Organization. (n.d.). https://www.fao.org/3/i2433e/i2433e02.pdf

[192]. Food and Agriculture Organization. "Chapter 5: Construction Materials". (n.d.). https://www.fao.org/3/i2433e/i2433e02.pdf

[193]. Los Almos National Laboratory. "LANL Sustainable Design Guide". Los Almos National Laboratory. 2022, December. https://engstandards.lanl.gov/esm/architectural/Sustainable.pdf

[194]. Los Almos National Laboratory. "LANL Sustainable Design Guide". Los Almos National Laboratory. 2022, December. https://engstandards.lanl.gov/esm/architectural/Sustainable.pdf

[195]. Active Sustainability. "Non-Polluting Construction Materials". Active Sustainability. (n.d.). https://www.activesustainability.com/construction-and-urban-development/non-polluting-construction-materials/?_adin=02021864894

[196]. Los Almos National Laboratory. "LANL Sustainable Design Guide". Los Almos National Laboratory. 2022, December. https://engstandards.lanl.gov/esm/architectural/Sustainable.pdf

[197]. Los Almos National Laboratory. "LANL Sustainable Design Guide". Los Almos National Laboratory. 2022, December. https://engstandards.lanl.gov/esm/architectural/Sustainable.pdf

[198]. Los Almos National Laboratory. "LANL Sustainable Design Guide". Los Almos National Laboratory. 2022, December. https://engstandards.lanl.gov/esm/architectural/Sustainable.pdf

[199]. Bellarmine. "Bamboo". Bellarmine. (n.d.). https://www.bellarmine.edu/faculty/drobinson/bamboo.asp

[200] Econation. "Bamboo". Econation. (n.d.). https://econation.one/bamboo/

[201]. Bambooder. Home page. Bambooder. (n.d.). https://www.bambooder.nl/

[202]. Kaiser, Cheryl. Ernst, Matt. "Bamboo". Centre for Crop Diversification Crop Profile. University of Kentucky. (n.d.). https://www.uky.edu/ccd/sites/www.uky.edu.ccd/files/bamboo.pdf

[203]. The United Nations - Department of Economic and Social Affairs. "Goal 11: Make cities and human settlements inclusive, safe, resilient and sustainable". Sustainable Development Goals. United Nations. (n.d.). https://sdgs.un.org/goals/goal11

[204]. UN-Habitat. "Rescuing SDG 11 for a Resilient Urban Planet". Synthesis Report. UN-Habitat. 2023. https://unhabitat.org/rescuing-sdg-11-for-a-resilient-urban-planet

[205]. United Nations Environment Programme. "Cities and Climate Change". Cities. United Nations Environment Programme. (n.d.). https://www.unep.org/explore-topics/resource-efficiency/what-we-do/cities/cities-and-climate-change

[206]. Basetti, Francesco. "The Cities of the Future". Climate Foresight. 2023, June 29. https://www.climateforesight.eu/articles/the-cities-of-the-future/

[207]. Dark Matter Labs. "Designing Our Future". Invitation Paper V01. Irresistible Circular Societies. Dark Matter Labs. 2023. https://www.irresistiblecircularsociety.eu/assets/uploads/20230707-New-European-Bauhaus-Economy_Digital-version_DML.pdf

[208]. Dark Matter Labs. "Designing Our Future". Invitation Paper V01. Desire. Irresistible Circular Societies. Dark Matter Labs. 2023. https://www.irresistiblecircularsociety.eu/assets/uploads/20230707-New-European-Bauhaus-Economy_Digital-version_DML.pdf

[209]. Dark Matter Labs. "Designing Our Future". Invitation Paper V01. Desire. Irresistible Circular Societies. Dark Matter Labs. 2023. https://www.irresistiblecircularsociety.eu/assets/uploads/20230707-New-European-Bauhaus-Economy_Digital-version_DML.pdf

[210] Dark Matter Labs. "Designing Our Future". Invitation Paper V01. Irresistible Circular Societies. Dark Matter Labs. 2023. https://www.irresistiblecircularsociety.eu/assets/uploads/20230707-New-European-Bauhaus-Economy_Digital-version_DML.pdf

[211]. Human-centred Artificial Intelligence. "Artificial Intelligence Definition". Stanford University. (n.d.). https://hai.stanford.edu/sites/default/files/2020-09/AI-Definitions-HAI.pdf

[212]. Smith, Chris. McGuire, Brian. Huang, Ting. Yang, Gary. "The History of Artificial Intelligence". History of Computing. University of Washington. 2006, December. https://courses.cs.washington.edu/courses/csep590/06au/projects/history-ai.pdf

[213]. Guerra, Abel. "The Future Benefits of Artificial Intelligence for Students". Artificial Intelligence. Urbe University. 2023, July 6. https://urbeuniversity.edu/blog/the-future-benefits-of-artificial-intelligence-for-students#:~:text=Automation%20of%20Administrative%20Tasks%3A%20AI, classes%2C%20and%20managing%20student%20records.

[214]. United Nations Climate Change. "AI for Climate Action: Technology Mechanism supports transformational climate solutions". News. United Nations Climate Change. 2023, November 3. https://unfccc.int/news/ai-for-climate-action-technology-mechanism-supports-transformational-climate-solutions

[215]. United Nations Climate Change. "AI for Climate Action: Technology Mechanism supports transformational climate solutions". News. United Nations Climate Change. 2023, November 3. https://unfccc.int/news/ai-for-climate-action-technology-mechanism-supports-transformational-climate-solutions

[216]. McLellan, Charles. "Fighting Climate Change: These 5 technologies are our best weapons". Sustainability. ZDNET. 2023, February 8.

https://www.zdnet.com/home-and-office/sustainability/fighting-climate-change-these-5-technologies-are-our-best-weapons/

[217]. Dark Matter Labs. Designing our Future". Invitation Paper V.0.1. Desire. Dark Matter Labs. 2023.https://www.irresistiblecircularsociety.eu/assets/uploads/20230707-New-European-Bauhaus-Economy_Digital-version_DML.pdf

[218]. European Parliament. "What is carbon neutrality and how can it be achieved by 2050?". News. European Parliament. 2019, October. https://www.europarl.europa.eu/news/en/headlines/society/20190926STO62270/what-is-carbon-neutrality-and-how-can-it-be-achieved-by-2050

[219]. Client Earth. "What is a Carbon Sink?". ClientEarth Communications. ClientEarth. 2020, December 22. https://www.clientEarth.org/latest/news/what-is-a-carbon-sink/

[220]. European Parliament. "What is carbon neutrality and how can it be achieved by 2050?". News. European Parliament. 2019, October. https://www.europarl.europa.eu/news/en/headlines/society/20190926STO62270/what-is-carbon-neutrality-and-how-can-it-be-achieved-by-2050

[221]. Dark Matter Labs. "Designing Our Future". A European Bauhaus Economy. Desire. Irresistible Circular Society. Dark Matter Labs. 2023, July 7. https://www.irresistiblecircularsociety.eu/assets/uploads/20230707-New-European-Bauhaus-Economy_Digital-version_DML.pdf

[222]. Solar Mag. "Solar Shingles: Turn Your Roof a Power Source (5 Brands).". Solar Roof. Solar Magazine. 2022, February 21. https://solarmagazine.com/solar-roofs/solar-shingles/

[223]. Verdinez, Deisy. "The Top 10 Countries for LEED demonstrate that green building is a truly global movement". Real Homes. 2023, February 7. https://www.realhomes.com/advice/types-of-eco-friendly-homes

[224]. Burton, Ann Loynd. "6 types of eco-friendly homes that create a more sustainable living space". Types of Eco-Friendly Homes. Real Homes. 2021, April 27. https://www.realhomes.com/advice/types-of-eco-friendly-homes

[225]. The Conversation. "The world needs to build more than two billion new homes over the next 80 years". The Conversation. 2018, February 28.

https://theconversation.com/the-world-needs-to-build-more-than-two-billion-new-homes-over-the-next-80-years-91794

[226]. Carbon Leadership Forums. "Life Cycle Assessment of Buildings". Carbon Leadership Forum. 2019. https://carbonleadershipforum.org/wp-content/uploads/2019/05/CLF-LCA-Practice-Guide_2019-05-23.pdf

[227]. Serhant, Ryan. "8 Eco-Friendly Features to Add to Your Home". Real Estate. Forbes. 2021, November 9. https://www.forbes.com/sites/ryanserhant/2021/11/09/8-eco-friendly-features-to-add-to-your-home/?sh=83724b1591dd

[228]. Energy Education. "Heat Pump". Encyclopedia. Energy Education. (n.d.). https://energyeducation.ca/encyclopedia/Heat_pump#:~:text=A%20heat%20pump%20is%20a,place%20to%20a%20warm%20place.

[229]. Thermodynamics Team B. "Heat Pump!!!". University of Wisconsin -Green Bay. 2019, November 27. https://blog.uwgb.edu/chem320b/heat-pump/

[230]. Gerhardt, Nick. Saddler, Lowe. "Heat Pump Vs. Air Conditioner: What Are The Differences?". Heat Pump Vs. Air Conditioner. Forbes. 2023, November 6. https://www.forbes.com/home-improvement/hvac/heat-pump-vs-air-conditioner/#:~:text=Heat%20pumps%20are%20more%20energy,twice%2Da%2Dyear%20maintenance.

[231]. European Parliament. "What is carbon neutrality and how can it be achieved by 2050?". News. European Parliament. 2019, October. https://www.europarl.europa.eu/news/en/headlines/society/20190926STO62270/what-is-carbon-neutrality-and-how-can-it-be-achieved-by-2050

[232]. GoodReads. "..Yes, but I have something he will never have … enough..". Quotable Quotes. (n.d.). https://www.goodreads.com/quotes/10651136-at-a-party-given-by-a-billionaire-on-shelter-island

[233]. Lee, Jenni. "7 Quotes on Climate Change and Health". Blog. United Nations Foundation. 2016, July 6. https://unfoundation.org/blog/post/7-quotes-on-climate-change-and-health/

[234]. The Institute for Health Metrics and Evaluation. "New Study Finds Poor Diet Kills More People Globally Than Tobacco and High Blood Pressure". News. Institute for Health Metrics and Evaluation. 2019, April 3. https://www.healthdata.org/news-events/newsroom/news-releases/new-

study-finds-poor-diet-kills-more-people-globally-tobacco-and#:~:text=Poor%20diet%20is%20responsible%20for,in%20every%20five%20deaths%20globally.

235. The Editors of Encyclopaedia Britannica. "Food". Arts & Culture. Britannica. 2023, October 29. https://www.britannica.com/topic/food

236. World Public Health Nutrition Association. "Food Classification. Public Health - Nova. The Star Shines Bright". World Nutrition - Volume 7. Number 1 - 3. World Public Health Nutrition Association. 2016, January - March. https://archive.wphna.org/wp-content/uploads/2016/01/WN-2016-7-1-3-28-38-Monteiro-Cannon-Levy-et-al-NOVA.pdf

237. World Health Organization, The. "Constitution". World Health Organization. (n.d). https://www.who.int/about/accountability/governance/constitution

238. Newsom, Rob. Rehman, Dr. Anis. "Diet, Exercise, and Sleep". Physical Health and Sleep. Sleep Foundation. 2023, March 3. https://www.sleepfoundation.org/physical-health/diet-exercise-sleep

239. Diary of a CEO. "The Junk Food Doctor: "THIS Food Is Worse Than Smoking!" - Chris Van Tulleken Ultra-Processed People". (15:20min)The Diary of a CEO. 2023, October 23. https://www.youtube.com/watch?v=dzUDhstqXbg

240. NHS Inform. "Malnutrition". Nutritional. NHS Inform. (n.d.). https://www.nhsinform.scot/illnesses-and-conditions/nutritional/malnutrition/

241. World Health Organization. "Obesity and Overweight". Detail. Fact Sheets. World Health Organization. 2021, June 9. https://www.who.int/news-room/fact-sheets/detail/obesity-and-overweight

242. Kluger, Jeffery. "More Than Half of the World Will Be Obese By 2035, Report Says". Health, Diet & Nutrition. Time. 2023, March 21. https://time.com/6264865/global-obesity-rates-increasing/

243. Center for Disease Control and Prevention. "Health Effects of Overweight & Obesity". Healthy Weight, Nutrition, and Physical Activity. Center for Disease Control and Prevention. (n.d.). https://www.cdc.gov/healthyweight/effects/index.html

244. Martinez, Prof. J. Alfredo. "Epigenetics within the double burden of malnutrition (Undernutrition and Obesity)". International Symposium on Understanding the Double Burden of Malnutrition for Effective Interventions. Universidad de Navarra. 2018. https://humanhealth.iaea.org/HHW/Nutrition/Symposium2018/presentations/7.2.Martinez.pdf

245. Morad, Renee. "How Diet Can Change Your DNA". Science for Life. Scientific American. 2017, June 7. https://www.scientificamerican.com/custom-media/science-for-life/how-diet-can-change-your-dna/

246. Good Reads. "Learning From History Quotes". Good Reads. (n.d.). https://www.goodreads.com/quotes/tag/learning-from-history

247. Föcking, Dorothea. "The Dutch Hunger Winter 1944-45". Keywords. Deutsches Museum München. Environment and Society. (n.d.). https://www.environmentandsociety.org/tools/keywords/dutch-hunger-winter-1944-45

248. University of Oxford. "What is the Food System". University of Oxford. (n.d.). https://www.futureoffood.ox.ac.uk/what-food-system

249. Tandon, Ayesha. "Food systems responsible for 'one third' of human-caused emissions". Carbon Brief. 2021, August 3. https://www.carbonbrief.org/food-systems-responsible-for-one-third-of-human-caused-emissions/

250. Tandon, Ayesha. "Food systems responsible for 'one third' of human-caused emissions". Carbon Brief. 2021, August 3. https://www.carbonbrief.org/food-systems-responsible-for-one-third-of-human-caused-emissions/

251. Karlsruher Institut für Technologie (KIT). "Expansion of agricultural land reduces carbon dioxide absorption". Science News. Science Daily. 2018, July 5. https://www.sciencedaily.com/releases/2018/07/180705115614.htm#:~:text=If%20forests%20are%20cut%20down,the%20atmosphere%2C%20plants%20and%20soil.

252. Karlsruher Institut für Technologie (KIT). "Expansion of agricultural land reduces carbon dioxide absorption". Science News. 2018, July 5. https://www.sciencedaily.com/releases/2018/07/180705115614.htm#:~:text=

If%20forests%20are%20cut%20down,the%20atmosphere%2C%20plants%20and%20soil.

[253]. Ritchie, Hannah. "Half of the world's habitable land is used for agriculture". Our World in Data. 2019, November 11. https://ourworldindata.org/global-land-for-agriculture

[254]. Ritchie, Hannah. "Half of the world's habitable land is used for agriculture". Our World in Data. 2019, November 11. https://ourworldindata.org/global-land-for-agriculture

[255]. Ritchie, Hannah. "Half of the world's habitable land is used for agriculture". Our World in Data. 2019, November 11. https://ourworldindata.org/global-land-for-agriculture

[256]. Milman, Oliver. "Meat accounts for nearly 60% of all greenhouse gases from food production, study finds". Eat Industry. The Guardian. 2021, September 13. https://www.theguardian.com/environment/2021/sep/13/meat-greenhouses-gases-food-production-study

[257]. Center For Sustainable Systems. "Carbon Footprint Factsheet". Factsheet. University of Michigan. (n.d.). https://css.umich.edu/publications/factsheets/sustainability-indicators/carbon-footprint-factsheet

[258]. Directorate-General for Environment. "Field to fork: global food miles generate nearly 20% of all CO2 emissions from food". News Article. European Commission. 2023, January 25. https://environment.ec.europa.eu/news/field-fork-global-food-miles-generate-nearly-20-all-co2-emissions-food-2023-01-25_en

[259]. Agriculture Victoria. "Livestock methane and nitrogen emissions". Understanding Carbon and Emissions. Agriculture Victoria. (n.d.). https://agriculture.vic.gov.au/climate-and-weather/understanding-carbon-and-emissions/livestock-methane-and-nitrogen-emissions#:~:text=Methane%20is%20the%20main%20greenhouse,that%20is%20then%20belched%20out.

[260]. Millar, Neville. Doll, Julie E., Robertson, G. Philip. "Management of Nitrogen Fertilizer to Reduce Nitrous Oxide (N2O) Emissions from Field Crops". Michigan State University. 2014, November.

https://www.canr.msu.edu/uploads/resources/pdfs/management_of_nitrogen_fertiler_(e3152).pdf

[261]. Lobell, David B. Tommaso, Stefania Di. Burney, Jennifer A. "Globally ubiquitous negative effects of nitrogen dioxide on crop growth". Environmental Studies. Science Advances. Science. 2022, June 1. https://www.science.org/doi/10.1126/sciadv.abm9909#:~:text=Nitrogen%20oxides%20(NOx)%20are,O3)%20and%20aerosol%20formation.

[262]. The National Institute for Occupational Safety and Health (NIOSH). "Nitrous Oxide". Center for Disease Control and Prevention. 2018, April 24. https://www.cdc.gov/niosh/topics/nitrousoxide/default.html#:~:text=Breathing%20nitrous%20oxide%20can%20cause,from%20exposure%20to%20nitrous%20oxide.

[263]. United Nations. "Food and Climate Change: Healthy diets for a healthier planet". Climate Action. United Nations. (n.d.). https://www.un.org/en/climatechange/science/climate-issues/food

[264]. National Geographic. "Diet". Encyclopedic Entry. Education. National Geographic. (n.d.). https://education.nationalgeographic.org/resource/diet/

[265]. Flatt Osborn, Jen. "How Many Vegans Are in the World? Exploring the Global Population of Vegans". Advocate. World Animal Foundation. 2023, September 16. https://worldanimalfoundation.org/advocate/how-many-vegans-are-in-the-world/#:~:text=Almost%201%E2%80%932%20Percent%20of,(The%20Vegan%20Society)

[266]. National Health Service. "The Vegan Diet". How to Eat a Balanced Diet. National Health Service. 2022, May 31. https://www.nhs.uk/live-well/eat-well/how-to-eat-a-balanced-diet/the-vegan-diet/

[267]. McManus, Katherine D. "What is a plant-based diet and why should you try it?". Harvard Health Blog. Harvard University. 2021, November 16. https://www.health.harvard.edu/blog/what-is-a-plant-based-diet-and-why-should-you-try-it-2018092614760

[268]. Kubala, Jillian. "Whole-Foods, Plant-Based Diet: A Detailed Beginner's Guide". Nutrition. HealthLine. 2023, June 23. https://www.healthline.com/nutrition/plant-based-diet-guide#overview

[269]. Richards, Louisa. "What is the difference between plant-based and vegan?". Articles. Medical News Today. 2021, April 9. https://www.medicalnewstoday.com/articles/plant-based-vs-vegan#key-differences

[270]. Hargreaves, Shila Minari. Rosenfeld, Daniel L. Moreira, , Ana Vládia Bandeira. Zandonad, Renata Puppin. "Plant-based and vegetarian diets: an overview and definition of these dietary patterns". National Library of Medicine. 2023, January 22. https://pubmed.ncbi.nlm.nih.gov/36681744/

[271]. Henderson, Lily. "Plant-based, vegetarian and vegan diets". Nutrition Facts. Heart Foundation. (n.d.). https://www.heartfoundation.org.nz/wellbeing/healthy-eating/nutrition-facts/plant-based-vegetarian-vegan-diets

[272]. Cleveland Clinic. "Mediterranean Diet". Health. Cleveland Clinic. 2022, November 20. https://my.clevelandclinic.org/health/articles/16037-mediterranean-diet

[273]. Cleveland Clinic. "Mediterranean Diet". Health. Cleveland Clinic. 2022, November 20. https://my.clevelandclinic.org/health/articles/16037-mediterranean-diet

[274]. Ajmera, Rachael. "What Is the Ayurvedic Diet? Benefits, Downsides, and More". Nutrition. Health Line. 2023, November 14. https://www.healthline.com/nutrition/ayurvedic-diet

[275]. Migala, Jessica. "5 Potential Health Benefits of an Ayurvedic Diet". Diet Nutrition. Everyday Health 2023, April 12. https://www.everydayhealth.com/diet-nutrition/potential-health-benefits-of-an-ayurvedic-diet/

[276]. Lad, Usha. Lad, Vasant. "Food Guidelines". Ayurveda. (n.d.). https://www.ayurveda.com/food-guidelines/

[277]. Hartfield, Emma. Torrens, Kerry. "What are ultra-processed foods?". Guide. BBC Good Food. 2023, June 8. https://www.bbcgoodfood.com/howto/guide/what-are-ultra-processed-foods

[278]. Monteiro, Carlos A. Cannon, Geoffrey. Levy,Renata B . Moubarac,Jean-Claude .Louzada, Maria LC. Rauber, Fernanda. Khandpur, Neha. Cediel, Gustavo. Neri, Daniela. Martinez-Steele, Euridice. Baraldi, Larissa G. Jaime,

Patricia C. "Ultra-processed foods: what they are and how to identify them". National Library of Medicine. 2019, April 12. https://www.ncbi.nlm.nih.gov/pmc/articles/PMC10260459/

[279]. Rauder, Fernanda. Martínez Steele, Eurídice. da Costa Louzada, Maria Laura. Millett, Christopher. Augusto Monteiro, Carlos. Bertazii Levy, Renata. Meyre, David. "Ultra-processed food consumption and indicators of obesity in the United Kingdom population (2008-2016)". PLoS One. 2020;15(5):e0232676. 2020, May 1. https://www.ncbi.nlm.nih.gov/pmc/articles/PMC7194406/

[280]. Thornton, Alex. "This is how many animals we eat each year". Food Security. World Economic Forum. 2019, February 8. https://www.weforum.org/agenda/2019/02/chart-of-the-day-this-is-how-many-animals-we-eat-each-year/

[281]. Hjálmarsdóttir, Freydís Guðný. "12 Foods That Are Very High in Omega-3". Nutrition. HealthLine. 2021, January 6. https://www.healthline.com/nutrition/12-omega-3-rich-foods#1.-Mackerel-(4,580-mg-per-serving)

[282]. Semeco, Arlene. "Top 12 Foods That Are High in Vitamin B12". Nutrition. HealthLine. 2023, January 23. https://www.healthline.com/nutrition/vitamin-b12-foods

[283]. CIEE. "Food - The Edible Part of Culture". University of Arkansas at Little Rock. (n.d.). https://ualr.edu/studyabroad/files/2012/05/CIEE_Food.pdf

[284]. Merriam-Webster. "Humane". Merriam-Webster Dictionary. 2023, November 29 (Updated). https://www.merriam-webster.com/dictionary/humane

[285]. Understanding Animal Research. "Cattle". (n.d.). https://www.understandinganimalresearch.org.uk/what-is-animal-research/a-z-animals/cattle

[286]. Animals Australia Team, "Inside the Secret Lives of Cows". Inspiring Stories. Animals Australia. 2023, May 3. https://animalsaustralia.org/our-work/inspiring-stories/secret-lives-of-cows/

[287]. Animals Australia Team, "Inside the Secret Lives of Cows". Inspiring Stories. Animals Australia Team. 2023, May 3. https://animalsaustralia.org/our-work/inspiring-stories/secret-lives-of-cows/

[288]. Animals Australia Team, "Inside the Secret Lives of Cows". Inspiring Stories. Animals Australia Team. 2023, May 3. https://animalsaustralia.org/our-work/inspiring-stories/secret-lives-of-cows/

[289]. Animal Welfare Institute. "Cattle". Animal Welfare Institute. (n.d.). https://awionline.org/content/cattle

[290]. Compassion in World Farming. "About Dairy Cows". Cows. (Compassion in World Farming. n.d.). https://www.ciwf.com/farmed-animals/cows/dairy-cows/#:~:text=Like%20humans%2C%20cows%20only%20produce,three%20months%20of%20giving%20birth.

[291]. Animal Welfare Institute. "Cattle". Animal Welfare Institute. (n.d.). https://awionline.org/content/cattle

[292]. Animal Welfare Institute. "Cattle". Animal Welfare Institute. (n.d.). https://awionline.org/content/cattle

[293]. Befoff, Marc. "The Emotional Lives and Personalities of Backyard Chickens". Relationships. Pyschology Today. 2023, March 26. https://www.psychologytoday.com/intl/blog/animal-emotions/202303/the-emotional-lives-and-personalities-of-backyard-chickens

[294]. Borkhataria, Cecile. "Never underestimate a chicken: Researchers find birds have distinct personalities, can count and even show Machiavellian-like tendencies". Science Tech. Daily Mail. 2017, January 3. https://www.dailymail.co.uk/sciencetech/article-4085004/Chickens-intelligent-empathy-distinct-personalities.html

[295]. Marino Lori. Thinking chickens: a review of cognition, emotion, and behaviour in the domestic chicken. Anim Cogn. 2017. National Library of Medicine. https://www.ncbi.nlm.nih.gov/pmc/articles/PMC5306232/

[296]. Royal Society for the Prevention of Cruelty to Animals. "Farming meat chickens". Farming. Royal Society for the Prevention of Cruelty to Animals. (n.d.). https://www.rspca.org.uk/adviceandwelfare/farm/meatchickens/farming

[297]. Royal Society for the Prevention of Cruelty to Animals. "Farming meat chickens". Farming. Royal Society for the Prevention of Cruelty to Animals. (n.d.). https://www.rspca.org.uk/adviceandwelfare/farm/meatchickens/farming

[298]. Lawrence, Felicity. "If consumers knew how farmed chickens were raised, they might never eat their meat again". The Observer. The Guardian. 2016, April 24. https://www.theguardian.com/environment/2016/apr/24/real-cost-of-roast-chicken-animal-welfare-farms

[299]. The Human League. "Factory-Farmed Chickens: The Cruelty of Chicken Farms". Chickens. The Human Society. 2021, January 19. https://thehumaneleague.org/article/factory-farmed-chickens

[300]. Lawrence, Felicity. "If consumers knew how farmed chickens were raised, they might never eat their meat again". The Observer. The Guardian. 2016, April 24. https://www.theguardian.com/environment/2016/apr/24/real-cost-of-roast-chicken-animal-welfare-farm

[301]. Watt, Nicholas. "Antimicrobial resistance a 'greater threat than cancer by 2050'". Antibiotics. The Guardian. 2016, April 14. https://www.theguardian.com/society/2016/apr/14/antimicrobial-resistance-greater-threat-cancer-2050-george-osborne

[302]. Watt, Nicholas. "Antimicrobial resistance a 'greater threat than cancer by 2050'". Antibiotics. The Guardian. 2016, April 14. https://www.theguardian.com/society/2016/apr/14/antimicrobial-resistance-greater-threat-cancer-2050-george-osborne

[303]. Watt, Nicholas. "Antimicrobial resistance a 'greater threat than cancer by 2050'". Antibiotics. The Guardian. 2016, April 14. https://www.theguardian.com/society/2016/apr/14/antimicrobial-resistance-greater-threat-cancer-2050-george-osborne

[304]. Brown, Culum. "Fish intelligence, sentience and ethics". Animal Cognition. Springer Link. 2014, June 19. https://link.springer.com/article/10.1007/s10071-014-0761-0

[305]. Stormberg, Joseph. "Are fish far more intelligent than we realize?". Vox. 2014, August 4. https://www.vox.com/2014/8/4/5958871/fish-intelligence-smart-research-behavior-pain

[306]. The Humane League. "Factory Fish Farming: What it is and Why It's Cruel to Fish". Animals. The Humane League. 2021, March 9. https://thehumaneleague.org/article/factory-fish-farming

[307]. The Humane League. "Factory Fish Farming: What it is and Why It's Cruel to Fish". Animals. The Humane League. 2021, March 9. https://thehumaneleague.org/article/factory-fish-farming

[308]. Watt, Nicholas. "Antimicrobial resistance a 'greater threat than cancer by 2050'". Antibiotics. The Guardian. 2016, April 14. https://www.theguardian.com/society/2016/apr/14/antimicrobial-resistance-greater-threat-cancer-2050-george-osborne

[309]. Meet the Press. "Bill Clinton; Sens. Webb & Kyl; Gov. Paterson on "Meet the Press". Real Clear Politics. 2009, September 27. Articles. https://www.realclearpolitics.com/articles/2009/09/27/bill_clinton_sens_webb__kyl_gov_paterson_on_meet_the_press_98485.html

[310]. BBC News. "The chicken that lived for 18 months without a head". BBC. 2015, September 10. https://www.bbc.com/news/magazine-34198390

[311]. Francione, Gary. "Animal welfare and society—Part 1, The viewpoints of a philosopher". Animal Frontiers. Oxford University Press. 2022, March 17. https://academic.oup.com/af/article/12/1/43/6550174

[312]. Kluger, Jeffery. "More Than Half of the World Will Be Obese By 2035, Report Says". Health, Diet & Nutrition. Time. 2023, March 21. https://time.com/6264865/global-obesity-rates-increasing/

[313]. Searing, Linda. "51 percent of world population may be overweight or obese by 2035". Big Number. 2023, March 20. https://www.washingtonpost.com/wellness/2023/03/20/obesity-overweight-increasing-worldwide/

[314]. World Health Organization. "World Obesity Day 2022 – Accelerating action to stop obesity". News Release. World Health Organization. 2022, March 4. https://www.who.int/news/item/04-03-2022-world-obesity-day-2022-accelerating-action-to-stop-obesity#:~:text=More%20than%201%20billion%20people,they%20are%20overweight%20or%20obese.

[315]. Omer, Sevil. "10 World Hunger Facts You Need to Know". In The Field. World Vision. 2023, August 28. https://www.worldvision.org/hunger-news-stories/world-hunger-facts#:~:text=As%20many%20as%20828%20million,to%20vulnerable%20populations%20and%20countries.

[316]. Omer, Sevil. "10 World Hunger Facts You Need to Know". In The Field. World Vision. 2023, August 28. https://www.worldvision.org/hunger-news-stories/world-hunger-facts#:~:text=As%20many%20as%20828%20million,to%20vulnerable%20populations%20and%20countries.

[317]. World Food Programme. "5 facts about food waste and hunger". Stories. World Food Programme. 2020, June 2. https://www.wfp.org/stories/5-facts-about-food-waste-and-hunger

[318]. World Food Programme. "5 facts about food waste and hunger". Stories. World Food Programme. 2020, June 2. https://www.wfp.org/stories/5-facts-about-food-waste-and-hunger

[319]. Global Goals. "Zero Hunger". Global Goals. (n.d.)., World Lead. https://www.globalgoals.org/goals/2-zero-hunger/

[320]. World Food Programme. "5 facts about food waste and hunger". Stories. World Food Programme. 2020, June 2. https://www.wfp.org/stories/5-facts-about-food-waste-and-hunger

[321]. Duram, Leslie A. "Organic Food". Arts & Culture. Britannica. 2023, November 23. https://www.britannica.com/topic/organic-food

[322]. WebMD Editorial Contributors. "What To Know About Obesogens". Weight Loss & Obesity. Web MD. 2023, August 10. https://www.webmd.com/obesity/what-to-know-obesogens

[323]. CBS News. "Chemicals in Food Can Make You Fat". The Early Show. CBS News. 2010, February 11. https://www.cbsnews.com/news/chemicals-in-food-can-make-you-fat/

[324]. Carrington, Damian. Arnett, George. "Clear differences between organic and non-organic food, study finds". Organic. The Guardian. 2014 July 11. https://www.theguardian.com/environment/2014/jul/11/organic-food-more-antioxidants-study

[325]. Klavinski, Rita. "7 Benefits of Eating Local Foods". News. College of Agriculture & Natural Resources. Michigan State University. 2013, April 13. https://www.canr.msu.edu/news/7_benefits_of_eating_local_foods

[326]. Directorate-General for Environment. "Field to fork: global food miles generate nearly 20% of all CO2 emissions from food". News Article. Environment. European Commission. 2023, January 25. https://environment.ec.europa.eu/news/field-fork-global-food-miles-generate-nearly-20-all-co2-emissions-food-2023-01-25_en

[327]. Directorate-General for Environment. "Field to fork: global food miles generate nearly 20% of all CO2 emissions from food". News Article. Environment. European Commission. 2023, January 25. https://environment.ec.europa.eu/news/field-fork-global-food-miles-generate-nearly-20-all-co2-emissions-food-2023-01-25_en

[328]. GoodReads. ""Avocado must be a magical fruit. The name itself sounds like an invocation." — Michael Bassey Johnson". Avocado Quotes. Quotes. Good Reads. (n.d.). https://www.goodreads.com/quotes/tag/avocado

[329]. Food Miles. "Food Miles Calculator - Tracked Item". Avocado. Food Miles. (n.d.). https://www.foodmiles.com/food/avocado

[330]. Usher, Tom. "This Is How Bad Your Avocado Obsession Is for the World". Article. Vice. 2018, May 25. https://www.vice.com/en/article/7xm8ab/this-is-how-bad-your-avocado-obsession-is-for-the-world

[331]. Jones, Benji. "The bad news about your avocado habit". Down to Earth. Vox. 2022, February 13. https://www.vox.com/down-to-Earth/22923403/super-bowl-avocados-mexico-deforestation

[332]. User, Tom. "This Is How Bad Your Avocado Obsession Is for the World". Environment. Vice.com. 2018, May 25. https://www.vice.com/en/article/7xm8ab/this-is-how-bad-your-avocado-obsession-is-for-the-world

[333]. ABC. "Spain: Avocado, the most fashionable fruit in Europe". Nieuwbericht. Agroberichten Buitenland. 2019, June 13. https://www.agroberichtenbuitenland.nl/actueel/nieuws/2019/06/13/spanish-avocados-in-europe

[334]. Swann, Asha. "This Genius Hack Keeps Avocados Perfectly Ripe for Months—Seriously". Blog. Brightly. 2021, November 19. https://brightly.eco/blog/how-to-store-avocado

[335]. Natural Resource Defense Council. "Regenerative Agriculture 101". Guide. Natural Resource Defense Council. 2021, November 29. https://www.nrdc.org/stories/regenerative-agriculture-101#what-is

[336]. Rainforest Alliance. "The Indigenous Roots of Regenerative Agriculture". Insights. Rainforest Alliance. 2023, August 9. https://www.rainforest-alliance.org/insights/the-indigenous-roots-of-regenerative-agriculture/

[337]. Rainforest Alliance. "The Indigenous Roots of Regenerative Agriculture". Insights. Rainforest Alliance. 2023, August 9. https://www.rainforest-alliance.org/insights/the-indigenous-roots-of-regenerative-agriculture/

[338]. Food and Agriculture Organization of the United Nations. "Agroforestry". Forestry. Food and Agriculture Organization. (n.d.). https://www.fao.org/forestry/agroforestry/80338/en/

[339]. Quandt, Amy. Neufeldt, Henry. Gorman, Kayla. "Climate change adaptation through agroforestry: opportunities and gaps". Current Opinion in Environmental Sustainability. Science Direct. 2023, February. https://www.sciencedirect.com/science/article/pii/S1877343522000963

[340]. USDA - Climate Hubs U.S. Department of Agriculture. "How Can Agroforestry Support Climate Change Mitigation in the Northeast". Northeast Climate Hub.USDA - Climate Hubs U.S. Department of Agriculture. (n.d.). https://www.climatehubs.usda.gov/hubs/northeast/topic/how-can-agroforestry-support-climate-change-mitigation-northeast#:~:text=Agroforestry%20contributes%20to%20climate%20change,and%20energy%20usage%20on%20farms.

[341]. SIWI. "Agroforestry for adaptation and mitigation to climate change". Policy Brief. Stockholm International Water Institute. 2020. https://siwi.org/publications/agroforestry-for-adaptation-and-mitigation-to-climate-change/

[342]. Spears, Stefanie. "What is No-Till Farming?". Regeneration International. 2018, June 6. https://regenerationinternational.org/2018/06/24/no-till-farming/

[343]. Cambridge Dictionary. "Tilling". Dictionary. Cambridge Dictionary. (n.d.). https://dictionary.cambridge.org/dictionary/english/tilling

[344]. Stewart, Robert E.. "Tillage". Agriculture. Science & Tech. Britannica. https://www.britannica.com/topic/tillage

[345]. Yao, Yao. Li, Guang. Lu, Yanhua. Liu, Shuainan. "Modelling the impact of climate change and tillage practices on soil CO2 emissions from dry farmland in the Loess Plateau of China". Ecological Modelling. Science Direct. 2023, February 4. https://www.sciencedirect.com/science/article/abs/pii/S0304380023000042#:~:text=Tillage%2C%20an%20important%20agricultural%20management,atmosphere%20(Bregaglio%20et%20al.%2C

[346]. Yao, Yao. Li, Guang. Lu, Yanhua. Liu, Shuainan. "Modelling the impact of climate change and tillage practices on soil CO2 emissions from dry farmland in the Loess Plateau of China". Ecological Modelling.Science Direct. 2023, February 4. https://www.sciencedirect.com/science/article/abs/pii/S0304380023000042#:~:text=Tillage%2C%20an%20important%20agricultural%20management,atmosphere%20(Bregaglio%20et%20al.%2C

[347]. World WildLife Foundation, Overview. World WildLife Foundation. (n.d.). https://www.worldwildlife.org/threats/soil-erosion-and-degradation

[348]. Creech, Elizabeth. "Saving Money, Time and Soil: The Economics of No-Till Farming". Conservation. Natural Resources Service. USDA. 2017, November 30. https://www.usda.gov/media/blog/2017/11/30/saving-money-time-and-soil-economics-no-till-farming

[349]. No Till Agriculture. "Advantages And Disadvantages Of No Till Farming". No-Till Farming. No Till Agriculture. (n.d.). https://notillagriculture.com/no-till-farming/advantages-and-disadvantages-of-no-till-farming/

[350]. NOVO Nordisk Foundation. "CO2 as a sustainable raw material in our future food production". News. NOVO Nordisk Foundation. 2023, June 13. https://novonordiskfonden.dk/en/news/co2-as-a-sustainable-raw-material-in-our-future-food-production/

[352]. NOVO Nordisk Foundation. "CO2 as a sustainable raw material in our future food production". News. NOVO Nordisk Foundation. 2023, June 13.

https://novonordiskfonden.dk/en/news/co2-as-a-sustainable-raw-material-in-our-future-food-production

353. Osher Center. "60% of chronic diseases could be prevented by a healthy diet". News. UW Department of Family Medicine. University of Washington. 2022, April 7. https://familymedicine.uw.edu/blog/cha-chi-ming-22-0407/#:~:text=It%20is%20estimated%20that%2060,inflammation%20that%20is%20already%20happening.

354. Gropper, Sareen S. "The Role of Nutrition in Chronic Disease". Nutrients. National Library of Medicine. 2023, February 15. https://www.ncbi.nlm.nih.gov/pmc/articles/PMC9921002/#:~:text=Diet%2C%20often%20considered%20as%20a,and%20perhaps%20some%20neurological%20diseases.

355. Tandon, Ayesha. "Food systems responsible for 'one third' of human-caused emissions". Carbon Brief. 2021, August 3. https://www.carbonbrief.org/food-systems-responsible-for-one-third-of-human-caused-emissions/

356. Bhandari, Smitha. "What is Dopamine?". Mental Health. 2023, July 19. WebMD. https://www.webmd.com/mental-health/what-is-dopamine

357. Yu, Yang. Miller, Renee. Groth, Susan W. Journal of Eating "A literature review of dopamine in binge eating". Journal of Eating Disorders. BMC Part of Sprinter Nature. 2022, January 28. https://jeatdisord.biomedcentral.com/articles/10.1186/s40337-022-00531-y#:~:text=The%20neurotransmitter%20dopamine%20is%20involved,and%20maintenance%20of%20binge%20eating.

358. Gunnars, Kris. "How to Identify and Manage Food Addiction" Nutrition. HealthLine. 2019, December 4. https://www.healthline.com/nutrition/how-food-addiction-works

359. Medical News Today. "Ultra-processed foods, especially artificial sweeteners, may increase depression risk". Articles. Medical News Today. (n.d.). https://www.medicalnewstoday.com/articles/ultra-processed-foods-may-be-as-addictive-as-smoking-study-says

360. Cleveland Clinic. "Meditation". Articles. Health. Cleveland Clinic. (n.d.). https://my.clevelandclinic.org/health/articles/17906-meditation

361. Katterman, Shawn N. Kleinman, Brighid M. Hood, Megan M. Nackers, Lisa M. Corsica, Joyce A. "Mindfulness meditation as an intervention for binge eating, emotional eating, and weight loss: A systematic review". Eating Behaviors. Elsevier. Science Direct. 2014, April 2. https://www.sciencedirect.com/science/article/abs/pii/S1471015314000191#:~:text=Results%20suggest%20that%20mindfulness%20meditation,term%20effects%20of%20mindfulness%20training.

362. Cleveland Clinic. "Hypnosis". Treatments. Cleveland Clinic. (n.d.). https://my.clevelandclinic.org/health/treatments/22676-hypnosis

363. Lindberg, Sara. "What to Know About Autogenic Training". Health. HealthLine. 2019, November 22. https://www.healthline.com/health/mental-health/autogenic-training

364. Tee-Melegrito, Rachel Ann. "Cortisol and stress: What is the connection?". Articles. Medical News Today. 2023, May 5. https://www.medicalnewstoday.com/articles/cortisol-and-stress

365. Jelinek, Joslyn. "Autogenic training: Benefits, limitations, and how to do it". Articles. Medical News Today. 2023, August 18. https://www.medicalnewstoday.com/articles/autogenic-training

366. Better Health Channel. "Placebo Effect". Health. Better Health Channel. (n.d.). https://www.betterhealth.vic.gov.au/health/conditionsandtreatments/placebo-effect

367. GoodReads. "Seeing is believing, but sometimes the most real things in the world are the things we can't see." — Chris Van Allsburg, The Polar Express". Good Reads. (n.d.). https://www.goodreads.com/quotes/8982769-seeing-is-believing-but-sometimes-the-most-real-things-in

368. Cheprasov, Artem. "Ignaz Semmelweis: Biography, Contribution to Medicine & Quotes". Study.com. (n.d.). https://study.com/academy/lesson/ignaz-semmelweis-biography-contribution-to-medicine-quotes.html

369. Leighton, Leslie S. "Ignaz Semmelweis, the doctor who discovered the disease-fighting power of hand-washing in 1847". The Conversation.com. 2020, April 14. https://theconversation.com/ignaz-semmelweis-the-doctor-who-discovered-the-disease-fighting-power-of-hand-washing-in-1847-135512#:~:text=We%20believe%20in%20the%20free%20flow%20of%20information&text=In%20fact%2C%20it%20was%2019th,prevent%20the%20spread%20of%20germs.

370. Leighton, Leslie S. "Ignaz Semmelweis, the doctor who discovered the disease-fighting power of hand-washing in 1847". The Conversation.com. 2020, April 14. https://theconversation.com/ignaz-semmelweis-the-doctor-who-discovered-the-disease-fighting-power-of-hand-washing-in-1847-135512#:~:text=We%20believe%20in%20the%20free%20flow%20of%20information&text=In%20fact%2C%20it%20was%2019th,prevent%20the%20spread%20of%20germs.

371. Williams, Sarah C.P. "Study Identifies Brain Areas Altered During Hypnotic Trances". What Hypnosis Does to the Brain.". Stanford Medical. Standford University. 2016, July 28. https://med.stanford.edu/news/all-news/2016/07/study-identifies-brain-areas-altered-during-hypnotic-trances.html

372. Hypnosis Training Academy. "Understanding The Powerful Link Between The Two". Hypnosis Training Academy. 2023, February 14. https://hypnosistrainingacademy.com/hypnosis-and-neuroscience/

373. Hypnosis Training Academy. "Achieving Your Fitness Goals With Hypnosis". Hypnosis Training Academy. 2023, November 6. https://hypnosistrainingacademy.com/weight-management-and-hypnosis/

374. Hypnosis Training Academy. "Self-Hypnosis (And Why The Super Successful Make Them Part of Their Morning Routines". Hypnosis Training Academy. 2023, March 31. https://hypnosistrainingacademy.com/difference-between-meditation-and-self-hypnosis/

375. Pilar Ramirex-Garcia, Maria. Leclerc-Loiselle, Jérôme. Genest, Christine. Lussier, Renaud.

376. Delagran, Louise. "What Are the Benefits of Mindfulness?". Mindfulness. Earl E. Bakken. University of Minnesota. (n.d.). https://www.takingcharge.csh.umn.edu/what-are-benefits-mindfulness

377. Global Goals. "Zero Hunger". Global Goals. (n.d.). https://www.globalgoals.org/goals/2-zero-hunger/

378. Global Goals. "Zero Hunger". Global Goals. (n.d.).https://www.globalgoals.org/goals/2-zero-hunger/

379. Montaño, A. Castro, A. de. "Pickling". Encyclopedia of Food and Health. Science Direct. 2016. https://www.sciencedirect.com/topics/food-science/pickling#:~:text=Pickling%20is%20the%20process%20of,the%20resulting%20foods%20as%20pickles.

380. Montaño, A. Castro, A. de. "Pickling". Encyclopedia of Food and Health. Science Direct. 2016. https://www.sciencedirect.com/topics/food-science/pickling#:~:text=Pickling%20is%20the%20process%20of,the%20resulting%20foods%20as%20pickles.

381. Treiber, Lisa. "Refrigerated pickled spring vegetables". News. Michigan State University Extension. Michigan State University. 2015, May 6. https://www.canr.msu.edu/news/refrigerated_pickled_spring_vegetables

382. Montaño, A. Castro, A. de. "Pickling". Encyclopedia of Food and Health. Science Direct. 2016. https://www.sciencedirect.com/topics/food-science/pickling#:~:text=Pickling%20is%20the%20process%20of,the%20resulting%20foods%20as%20pickles.

383. Global Goals. "Zero Hunger". Global Goals. (n.d.).https://www.globalgoals.org/goals/2-zero-hunger/

384. Huang, Jiaqi. M. Liao, Linda. J. Weinstein, Stephanie. Sinha, Rashmi. I. Grauhard, Barry. Albanas, Demetrius. "Association Between Plant and Animal Protein Intake and Overall and Cause-Specific Mortality". National Library of Medicine. 2020, September 1.

https://pubmed.ncbi.nlm.nih.gov/32658243/#:~:text=Replacement%20of%203%25%20energy%20from,%25%20lower%20risk%20in%20women).

385. PETA. "11 Vegan Foods That Are Complete Protein Sources", Food & Health. PETA. 2017, July 26. https://www.peta.org/living/food/complete-proteins-vegan/

386. Weishaupt, Jeffrey. "What Are Plant Food Sources of Vitamin B12?". Diet & Weight Management / Reference. WebMD. 2021, November 24. https://www.webmd.com/diet/what-are-plant-food-sources-vitamin-b12

387. Safi, Michael. "Angela Merkel pressures Australia to reveal its greenhouse gas targets". The Guardian. 2014, November 14. https://www.theguardian.com/world/2014/nov/17/angela-merkel-pressures-australia-to-reveal-its-greenhouse-gas-targets

388. Klein Vision. "The flying car completes first ever inter-city flight (Official Video)". YouTube. 2021, June 29. https://www.youtube.com/watch?v=a2tDOYkFCYo

389. ELTIS. "Mobility". Glossary. ELTIS. 2019, May 28. https://www.eltis.org/glossary/mobility

390. Reid, Carlton. "You Are Not Stuck In Traffic, You Are Traffic". Sustainability. Forbes. 2018, December 3. https://www.forbes.com/sites/carltonreid/2018/12/03/you-are-not-stuck-in-traffic-you-are-traffic/?sh=22501bd4583d

391. Hewitt, William. "The Bleichert system of aerial tramways. Reversible aerial tramways. Aerial tramways of special design". University of California. 1909. https://archive.org/details/bleichertsystem01hewigoog

392. National Aeronautics and Space Administration. "Luna 1". NASA. (n.d.). https://nssdc.gsfc.nasa.gov/nmc/spacecraft/display.action?id=1959-012A

393. Mihlfeld & Associates. "The 6 Modes of Transportation". Blog. Mihfeld & Associates. 2018, October 19. https://blog.mihlfeld.com/the-6-modes-of-transportation#:~:text=Therefore%3B%20an%20essential%20part%20of,rail%2C%20intermodal%2C%20and%20pipeline.

394. International Transport Forum. "Measuring New Mobility: Definitions, Indicators, Data Collection". OECD/ITF. 2023, May 9. https://www.itf-oecd.org/sites/default/files/docs/measuring-new-mobility-definitions-indicators-data.pdf

395. Millian, Laura. "Switching From Cars to Bikes Cuts Commuting Emissions by 67%". Energy and Science. Bloomberg. 2021, March 31. https://www.bloomberg.com/news/articles/2021-03-31/switching-from-cars-to-bikes-cuts-commuting-emissions-by-67#xj4y7vzkg

396. Ryan, Andrew. "Powered Light Vehicles 'Essential for Future Urban Mobility". Mobility Strategy. FleetNews. 2019, December 13.

https://www.fleetnews.co.uk/fleet-management/future-fleet/powered-light-vehicles-essential-for-future-urban-mobility

[397]. Lafayette Proctor, Charles. "Internal combustion engine". Britannica. 2023, August 18. https://www.britannica.com/technology/internal-combustion-engine

[398]. International Transport Forum. "Measuring New Mobility: Definitions, Indicators, Data Collection". OECD/ITF. 2023, May 9. https://www.itf-oecd.org/sites/default/files/docs/measuring-new-mobility-definitions-indicators-data.pdf

[399]. KIA. "What is a Hybrid Car?". KIA. (n.d.). https://www.kia.com/mu/discover-kia/ask/what-is-a-hybrid-car.html

[400]. International Transport Forum. "Measuring New Mobility: Definitions, Indicators, Data Collection". OECD/ITF. 2023, May 9. https://www.itf-oecd.org/sites/default/files/docs/measuring-new-mobility-definitions-indicators-data.pdf

[401]. Cole, Rachel. "Autonomous Vehicle". Britannica. 2023, October 23. https://www.britannica.com/technology/autonomous-vehicle

[402]. International Transport Forum. "Measuring New Mobility: Definitions, Indicators, Data Collection". OECD/ITF. 2023, May 9. https://www.itf-oecd.org/sites/default/files/docs/measuring-new-mobility-definitions-indicators-data.pdf

[403]. Heineke, Kersten. Laverty, Nicholas. Möller, Timo. Ziegler, Felix. "The Future of Mobility". Industries. Mckinsey Quarterly. McKinsey. 2023, April 19. https://www.mckinsey.com/industries/automotive-and-assembly/our-insights/the-future-of-mobility-mobility-evolves

[404]. International Energy Agency. "Transport". Energy System. International Energy Agency. 2023, July 11. https://www.iea.org/energy-system/transport

[405]. Siemens Xcelerator. "Mobile apps - your digital companion". Siemens. 2023. https://www.mobility.siemens.com/global/en/portfolio/intermodal/apps.html

[406]. International Transport Forum. "Measuring New Mobility: Definitions, Indicators, Data Collection". OECD/ITF. 2023, May 9. https://www.itf-oecd.org/sites/default/files/docs/measuring-new-mobility-definitions-indicators-data.pdf

[407]. Neild, Barry. "Which City has the World's Best Taxis?". Articles. CNN. 2017, July 13. https://edition.cnn.com/travel/article/worlds-best-taxis/index.html

[408]. Bilgin, Pinar. Mattioli, Giulio. Morgan, Malcolm. Wadud, Zia. "The effects of ridesourcing services on vehicle ownership: The case of Great Britain".

University of Leeds; TU Dortmund University. 2022, September 19. https://www.sciencedirect.com/science/article/pii/S1361920923000718

409. Silva, Dave. "What is Microtransit?". Blog. TripSpark.com. (n.d.). https://www.tripspark.com/blog/what-is-microtransit-definition-benefits/

410. Urbanism Next. "Microtransit". Technologies. Urbanism Next. (n.d.). https://www.urbanismnext.org/technologies/microtransit

411. MOIA. "Ridepooling: What is it and what does it do?". Blog. MOIA. (n.d.). https://www.moia.io/en/blog/ridepooling#:~:text=Ridepooling%20is%20a%20digital%2Dbased,carried%20out%20by%20an%20algorithm.

412. Zwick, Felix. Axhausen, Kay W. "Ride-pooling demand prediction: A spatiotemporal assessment in Germany". Journal of Transport Geography. Elsevier. Science Direct. 2022, April. https://www.sciencedirect.com/science/article/pii/S0966692322000308

413. Law Insider. "Share Fleet Vehicles". Dictionary. Law Insider. (n.d.). https://www.lawinsider.com/dictionary/shared-fleet-vehicles#:~:text=Shared%20fleet%20vehicles%20means%20any,use%20over%20discrete%20time%20intervals.

414. International Transport Forum. "Measuring New Mobility: Definitions, Indicators, Data Collection". OECD/ITF. 2023, May 9. https://www.itf-oecd.org/sites/default/files/docs/measuring-new-mobility-definitions-indicators-data.pdf

415. Rubiano, Leonardo Canon. "Zero docks: what we learnt about dockless bike-sharing during #TTDC2018". Transport. Transport for Development. World Bank. 2018, January 25. https://blogs.worldbank.org/transport/zero-docks-what-we-learnt-about-dockless-bike-sharing-during-ttdc2018

416. Rubiano, Leonardo Canon. "Zero docks: what we learnt about dockless bike-sharing during #TTDC2018". Transport. Transport for Development. World Bank. 2018, January 25. https://blogs.worldbank.org/transport/zero-docks-what-we-learnt-about-dockless-bike-sharing-during-ttdc2018

417. International Transport Forum. "Measuring New Mobility: Definitions, Indicators, Data Collection". OECD/ITF. 2023, May 9. https://www.itf-oecd.org/sites/default/files/docs/measuring-new-mobility-definitions-indicators-data.pdf

418. Wunder Mobility. "What is a car sharing service?". Blog. 2022, August 29. https://www.wundermobility.com/blog/what-is-a-car-sharing-service#:~:text=Per%20definition%2C%20car%20sharing%20is,hours%20or%20pick%2Dup%20locations.

419. FleetNews. "Nine things you need to know about corporate car sharing". Rental. Fleet News. 2017, February 23. https://www.fleetnews.co.uk/fleet-management/rental/nine-things-you-need-to-know-about-corporate-car-sharing

[420]. International Transport Forum. "Measuring New Mobility: Definitions, Indicators, Data Collection". OECD/ITF. 2023, May 9. https://www.itf-oecd.org/sites/default/files/docs/measuring-new-mobility-definitions-indicators-data.pdf

[421]. GlobalData. "Top 10 Car Rental Companies in the World in 2021 by Revenue". Travel and Tourism. Global Data. (n.d.). https://www.globaldata.com/companies/top-companies-by-sector/travel-and-tourism/global-car-rental-companies-by-revenue/

[422]. International Transport Forum. "Measuring New Mobility: Definitions, Indicators, Data Collection". OECD/ITF. 2023, May 9. https://www.itf-oecd.org/sites/default/files/docs/measuring-new-mobility-definitions-indicators-data.pdf

[423]. International Transport Forum. "Measuring New Mobility: Definitions, Indicators, Data Collection". OECD/ITF. 2023, May 9. https://www.itf-oecd.org/sites/default/files/docs/measuring-new-mobility-definitions-indicators-data.pdf

[424]. Dictionary. "Jitney". (n.d.). https://www.dictionary.com/browse/jitney

[425]. International Transport Forum. "Measuring New Mobility: Definitions, Indicators, Data Collection". OECD/ITF. 2023, May 9. https://www.itf-oecd.org/sites/default/files/docs/measuring-new-mobility-definitions-indicators-data.pdf

[426]. International Transport Forum. "Measuring New Mobility: Definitions, Indicators, Data Collection". OECD/ITF. 2023, May 9. https://www.itf-oecd.org/sites/default/files/docs/measuring-new-mobility-definitions-indicators-data.pdf

[427]. Law Insider. "Urban Development definition". Dictionary. Law Insider. (n.d.). https://www.lawinsider.com/dictionary/urban-development

[428]. Heineke, Kersten. Laverty, Nicholas. Möller, Timo. Ziegler, Felix. "The Future of Mobility". Industries. Mckinsey Quarterly. McKinsey. 2023, April 19. https://www.mckinsey.com/industries/automotive-and-assembly/our-insights/the-future-of-mobility-mobility-evolves

[429]. Irwin-Hunt, Alex. "Traffic Time: The World's Most Congested Cities". Data Trends. FDI Intelligence. 2023, June 28.

https://www.fdiintelligence.com/content/data-trends/traffic-time-the-worlds-most-congested-cities-82680

[430]. Irwin-Hunt, Alex. "Traffic Time: The World's Most Congested Cities". Data Trends. FDI Intelligence. 2023, June 28. https://www.fdiintelligence.com/content/data-trends/traffic-time-the-worlds-most-congested-cities-82680

[431]. Morris, David. Z. "Today's Cars Are Parked 95% of the Time". Tech. Transportation. Fortune. 2016, March 13. https://fortune.com/2016/03/13/cars-parked-95-percent-of-time/

[432]. Heineke, Kersten. Laverty, Nicholas. Möller, Timo. Ziegler, Felix. "The Future of Mobility". Industries. Mckinsey Quarterly. McKinsey. 2023, April 19. https://www.mckinsey.com/industries/automotive-and-assembly/our-insights/The-future-of-mobility-global-implications

[433]. Heineke, Kersten. Laverty, Nicholas. Möller, Timo. Ziegler, Felix. "The Future of Mobility". Industries. Mckinsey Quarterly. McKinsey. 2023, April 19. https://www.mckinsey.com/industries/automotive-and-assembly/our-insights/The-future-of-mobility-global-implications

[434]. McKerracher, Colin. "The US Could Become the Odd Market Out in the EV Success Story". Newsletter - Hyperdrive. Bloomberg. 2023, October 30. https://www.bloomberg.com/news/newsletters/2023-10-30/the-us-could-become-the-odd-market-out-in-the-ev-success-story?cmpid=BBD103023_hyperdrive&utm_medium=email&utm_source=newsletter&utm_term=231030&utm_campaign=hyperdrive

[435]. Heineke, Kersten. Laverty, Nicholas. Möller, Timo. Ziegler, Felix. "The Future of Mobility". Industries. Mckinsey Quarterly. McKinsey. 2023, April 19. https://www.mckinsey.com/industries/automotive-and-assembly/our-insights/The-future-of-mobility-global-implications

[436]. Rhodes, Anna. "Uber: Which countries have banned the controversial taxi app". News & Advice. Independent. 2017, September 22. https://www.independent.co.uk/travel/news-and-advice/uber-ban-countries-where-world-taxi-app-europe-taxi-us-states-china-asia-legal-a7707436.html

[437]. Davis, Levi. "Lyft vs Uber: What's the Difference?". Investopedia. 2023, May 26. https://www.investopedia.com/articles/personal-finance/010715/key-differences-between-uber-and-lyft.asp

[438]. Heineke, Kersten. Laverty, Nicholas. Möller, Timo. Ziegler, Felix. "The Future of Mobility". Industries. Mckinsey Quarterly. McKinsey. 2023, April 19. https://www.mckinsey.com/industries/automotive-and-assembly/our-insights/The-future-of-mobility-global-implications

[439]. Heineke, Kersten. Laverty, Nicholas. Möller, Timo. Ziegler, Felix. "The Future of Mobility". Industries. Mckinsey Quarterly. McKinsey. 2023, April 19. https://www.mckinsey.com/industries/automotive-and-assembly/our-insights/The-future-of-mobility-global-implications

[440]. Cole, Rachel. "Autonomous Vehicles". Science & Tech. Britannica. 2023, October 3. https://www.britannica.com/technology/autonomous-vehicle

[441]. Synopsys. "The 6 Levels of Vehicle Autonomy Explained". Synopsys. (n.d.). https://www.synopsys.com/automotive/autonomous-driving-levels.html

[442]. Synopsys. "The 6 Levels of Vehicle Autonomy Explained". Synopsys. (n.d.). https://www.synopsys.com/automotive/autonomous-driving-levels.html

[443]. Dow, Cat. "What are the Six SAE Levels of Self-driving Cars?". Top Gear Advice. 2023, March 6. https://www.topgear.com/car%20news/what-are-sae-levels-autonomous-driving-uk

[444]. Rambus. "SAE Levels of Automation in Cars Simply Explained (+image)". Rambus Press. 2022, June 9. https://www.rambus.com/blogs/driving-automation-levels/#:~:text=lane%20departure%20warning.-,Level%201%3A%20Driver%20assistance,control%20and%20lane%20keeping%20assistance.

[445]. Bogna, John. "Is Your Car Autonomous? The 6 Levels of Self-Driving Explained". NextCar. PC Mag. 2022, June 14. https://www.pcmag.com/how-to/6-levels-of-autonomous-self-driving-explained

[446]. Bogna, John. "Is Your Car Autonomous? The 6 Levels of Self-Driving Explained". NextCar. PC Mag. 2022, June 14. https://www.pcmag.com/how-to/6-levels-of-autonomous-self-driving-explained

[447]. Bogna, John. "Is Your Car Autonomous? The 6 Levels of Self-Driving Explained". NextCar. 2022, June 14. https://www.pcmag.com/how-to/6-levels-of-autonomous-self-driving-explained

[448]. Golson, Jordan. Bohn, Dieter. "All new Tesla cars now have hardware for 'full self-driving capabilities' / But some safety features will be disabled initially". Tesla. The Verge. 2016, October 20. https://www.theverge.com/2016/10/19/13340938/tesla-autopilot-update-model-3-elon-musk-update

[449]. College of Engineering. "Pros/Cons". Autonomous Systems. The Ohio State University. (n.d.). https://u.osu.edu/autonomousvehicles/proscons/

[450]. College of Engineering. "Pros/Cons". Autonomous Systems. The Ohio State University. (n.d.). https://u.osu.edu/autonomousvehicles/proscons/

[451]. College of Engineering. "Pros/Cons". Autonomous Systems. The Ohio State University. (n.d.). https://u.osu.edu/autonomousvehicles/proscons/

[452]. World Health Organization. "Road Traffic Injuries". Details. Fact Sheet. Newsroom. 2022, June 20. https://www.who.int/news-room/fact-sheets/detail/road-traffic-injuries

[453]. Harvard Business School. "Business Insights. Harvard Business School Online. 2021` June 8. https://online.hbs.edu/blog/post/why-is-gdp-important

[454]. IRU Intelligence Briefing. "Driver Shortage Global Report 2022". https://unece.org/sites/default/files/2022-09/WP5_Session35_Agenda8a_Marie-Anne%20Cervoni.pdf

[455]. LeVine, Steve. "What it really costs to turn a car into a self-driving vehicle". Tech & Innovation. 2017, March 5. https://qz.com/924212/what-it-really-costs-to-turn-a-car-into-a-self-driving-vehicle

[456]. Mckinsey. "Autonomous driving's future: Convenient and connected". Report. McKinsey. 2023, January 6. https://www.mckinsey.com/industries/automotive-and-assembly/our-insights/autonomous-drivings-future-convenient-and-connected

[457]. Winton, Neil. Computer Driven Autos Still Years Away Despite Massive Investment". Transportation. Forbes. 2022, February 27. https://www.forbes.com/sites/neilwinton/2022/02/27/computer-driven-autos-still-years-away-despite-massive-investment/?sh=3b15087a18cc

[458]. Winton, Neil. Computer Driven Autos Still Years Away Despite Massive Investment". Transportation. Forbes. 2022, February 27.

https://www.forbes.com/sites/neilwinton/2022/02/27/computer-driven-autos-still-years-away-despite-massive-investment/?sh=3b15087a18cc

[459]. Cambridge Dictionary. "Cargo Bike". Dictionary. Cambridge. (n.d.). https://dictionary.cambridge.org/dictionary/english/cargo-bike

[460]. Cambridge Dictionary. "Pedicab". Dictionary. Cambridge. (n.d.). https://dictionary.cambridge.org/dictionary/english/pedicab

[461]. Merriam-Webster. "Electric Bike". Dictionary. Cambridge. (n.d.). https://www.merriam-webster.com/dictionary/electric%20bike

[462]. Cambridge Dictionary. "Pedelec". Dictionary. Cambridge. (n.d.). https://dictionary.cambridge.org/dictionary/english/pedelec

[463]. Cambridge Dictionary. "Speed pedelec". Dictionary. Cambridge. (n.d.). https://dictionary.cambridge.org/dictionary/english/speed-pedelec

[464]. International Transport Forum. "Measuring New Mobility: Definitions, Indicators, Data Collection". OECD/ITF. 2023, May 9. https://www.itf-oecd.org/sites/default/files/docs/measuring-new-mobility-definitions-indicators-data.pdf

[465]. Merriam-Webster. "Scooter". Dictionary. Merriam-Webster. (n.d.). https://www.merriam-webster.com/dictionary/scooter#:~:text=%3A%20a%20vehicle%20ridden%20while%20standing,make%20today's%20scooters%20relatively%20safe.

[466]. Merriam-Webster. "E-Scooter". Dictionary. Merriam-Webster. (n.d.). https://www.merriam-webster.com/dictionary/e-scooter

[467]. Merriam-Webster. "Mobility Scooter". Dictionary. Merriam-Webster. (n.d.). https://www.merriam-webster.com/dictionary/mobility%20scooter

[468]. International Transport Forum. "Measuring New Mobility: Definitions, Indicators, Data Collection". OECD/ITF. 2023, May 9. https://www.itf-oecd.org/sites/default/files/docs/measuring-new-mobility-definitions-indicators-data.pdf

[469]. Merriam-Webster. "Skateboard". Dictionary. Merriam-Webster. (n.d). https://www.merriam-webster.com/dictionary/skateboard

[470]. International Transport Forum. "Measuring New Mobility: Definitions, Indicators, Data Collection". OECD/ITF. 2023, May 9. https://www.itf-oecd.org/sites/default/files/docs/measuring-new-mobility-definitions-indicators-data.pdf

[471]. International Transport Forum. "Measuring New Mobility: Definitions, Indicators, Data Collection". OECD/ITF. 2023, May 9. https://www.itf-oecd.org/sites/default/files/docs/measuring-new-mobility-definitions-indicators-data.pdf

[472]. International Transport Forum. "Measuring New Mobility: Definitions, Indicators, Data Collection". OECD/ITF. 2023, May 9. https://www.itf-oecd.org/sites/default/files/docs/measuring-new-mobility-definitions-indicators-data.pdf

[473]. Cambridge Dictionary. "Hoverboard". Dictionary. Cambridge. (n.d.). https://dictionary.cambridge.org/dictionary/english/hoverboard

[474]. Brahambhatt, Rupendra. "Hoverboards are now real — and the science behind them is dope". Inventions. ZME Science. 2023, May 4. https://www.zmescience.com/feature-post/technology-articles/inventions-1/hoverboards-real-science-07112021

[475]. Hubert, "The Benefits of Micro-mobility for Cities"., News. ELTIS. 2021, December 6. https://www.eltis.org/in-brief/news/benefits-micro-mobility-cities

[476]. TomTom. "Ranking 2022". Traffic-Index. TomTom. (n.d.). https://www.tomtom.com/traffic-index/ranking/

[477]. Beedham, Matthew. "What does TomTom Traffic Index data tell us about the world's busiest cities?". Explainers And Insights. TomTom. 2023, February 15. https://www.tomtom.com/newsroom/explainers-and-insights/the-most-congested-cities-in-the-world-2022/

[478]. Hubert, "The Benefits of Micro-mobility for Cities"., News. ELTIS. 2021, December 6. https://www.eltis.org/in-brief/news/benefits-micro-mobility-cities

[479]. Hubert, "The Benefits of Micro-mobility for Cities"., News. ELTIS. 2021, December 6. https://www.eltis.org/in-brief/news/benefits-micro-mobility-cities

480. Hubert, "The Benefits of Micro-mobility for Cities"., News. ELTIS. 2021, December 6. https://www.eltis.org/in-brief/news/benefits-micro-mobility-cities

481. Kolata Gina. "The Bicycling Paradox: Fit Doesn't Have to Mean Thin". Fitness. New York Times. 2007, July 17. https://www.nytimes.com/2007/07/17/health/nutrition/17essa.html

482. TATA AIG Team. "Highway Hypnosis". Car Insurance. TATA AIG. 2022, September 13. https://www.tataaig.com/knowledge-center/car-insurance/highway-hypnosis

483. Puttkamer, Laura. "These are the world's most bike-friendly cities". Society. TOPOS Magazine. 2023, January 9. https://toposmagazine.com/bike-friendly-cities/

484. FitzPatrick, Alex. "These cities have the most bicycle deaths per capita". Health. Axios. 2023, May 18. https://www.axios.com/2023/05/18/bike-deaths-by-city

485. Wang, Angela. "Cyclist deaths are rising nationwide — here are the 20 most dangerous cities to ride a bike in". Travel. Insider. 2019, July 19. https://www.insider.com/most-dangerous-cities-to-ride-a-bike-in-2019-7

486. Ericsson. "Micromobility Challenges". Shared challenges for micromobility operators. Ericsson. (n.d.). https://www.ericsson.com/en/enterprise/reports/connected-micromobility/pain-points

487. Ericsson. "Micromobility Challenges". Shared challenges for micromobility operators. Ericsson. (n.d.). https://www.ericsson.com/en/enterprise/reports/connected-micromobility/pain-points

488. International Transport Forum. "Measuring New Mobility: Definitions, Indicators, Data Collection". OECD/ITF. 2023, May 9. https://www.itf-oecd.org/sites/default/files/docs/measuring-new-mobility-definitions-indicators-data.pdf

489. International Transport Forum. "Measuring New Mobility: Definitions, Indicators, Data Collection". OECD/ITF. 2023, May 9. https://www.itf-oecd.org/sites/default/files/docs/measuring-new-mobility-definitions-indicators-data.pdf

490. Merriam-Webster. "Moped". Dictionary. (n.d.). https://www.merriam-webster.com/dictionary/moped

491. International Transport Forum. "Measuring New Mobility: Definitions, Indicators, Data Collection". OECD/ITF. 2023, May 9. https://www.itf-oecd.org/sites/default/files/docs/measuring-new-mobility-definitions-indicators-data.pdf

[492]. International Transport Forum. "Measuring New Mobility: Definitions, Indicators, Data Collection". OECD/ITF. 2023, May 9. https://www.itf-oecd.org/sites/default/files/docs/measuring-new-mobility-definitions-indicators-data.pdf

[493] Jelbi. "Jelbi". (n.d.). https://www.jelbi.de/en/home/

[494]. Tracxn. "Competitive Landscape of Jelbi". Tracxn. 2023, October 7. https://tracxn.com/d/companies/jelbi/__uuA_KsCleFS26UndaI2t3Qj0dVyt4wz8arZfnyhqrT0/competitors

[495]. Jenkinson, Lloyd R. Marchman III, James F. "7 - Project study: a dual-mode (road/air) vehicle". Aircraft Design Project for Engineering Students. Science Direct. 2007, September 2. https://www.sciencedirect.com/science/article/abs/pii/B978075065772350090

[496]. The Economist. "A flying car that anyone can use will soon go on sale". Science and Technology. Economist. 2023, October 10. https://www.economist.com/science-and-technology/2023/10/10/a-flying-car-that-anyone-can-use-will-soon-go-on-sale

[497] Euronews and AP. "This US start-up is set to begin test flights of their 'flying car' prototype". Mobility. Euro News. 2023, September 22. https://www.euronews.com/next/2023/09/22/this-us-start-up-is-set-to-begin-test-flights-of-their-flying-car-prototype

[498] . Tornatore, Cinzia. Marchitto, Luca. Sabia, Pino. Joannon, Mara De. "Ammonia as Green Fuel in Internal Combustion Engines: State-of-the-Art and Future Perspectives". Insights in Engine and Automotive Engineering: Frontiers. 2021. 2021, July, 22. https://www.frontiersin.org/articles/10.3389/fmech.2022.944201/full

[499]. Lew, Linda. "China's GAC Unveils World's First Ammonia Car Engine". Hyperdrive. Bloomberg. 2023, June 26. https://www.bloomberg.com/news/articles/2023-06-26/china-s-gac-unveils-world-s-first-ammonia-car-engine

[500]. Anderson, Brad. "GAC And Toyota Have Created An Ammonia-Powered 2.0-Liter Engine". Tech. CarsCoops. 2023, July 5. https://www.carscoops.com/2023/07/gac-and-toyota-have-created-an-ammonia-powered-2-0-liter-engine/

[501]. Jaeger, Joel. "These Countries Are Adopting Electric Vehicles the Fastest". Insights. The World Resources Institute. 2023, September 14. https://www.wri.org/insights/countries-adopting-electric-vehicles-fastest#:~:text=The%20top%205%20countries%20with,%25)%2C%20according%20to%20our%20analysis.

[502]. Worldometer. "Regions in the world by population (2023)". World. Worldometer. https://www.worldometers.info/world-population/population-by-region/

503. Dankwa Ampah, Jeffrey. Afrane, Sandylove. Bonah Agyekum, Ephraim. Adun, Humphrey. Abdu Yusuf, Abdulfatah. Bamisile, "Electric vehicles development in Sub-Saharan Africa: Performance assessment of standalone renewable energy systems for hydrogen refuelling and electricity charging stations (HRECS)". Journal of Cleaner Production. Science Direct. 2022, November 20. https://www.sciencedirect.com/science/article/abs/pii/S0959652622038100

504. UN Environment Programme. "In face of rising air pollution, Rwanda turns to electric vehicles". Story. UN Environment Programme. 2022, October 31. https://www.unep.org/news-and-stories/story/face-rising-air-pollution-rwanda-turns-electric-vehicles#:~:text=Close%20to%20900%20locally%20made,first%20electric%20Volkswagen%20in%20Africa.

505. Whiting, Kate. "Africa's motorbike taxis are going electric - saving money and cutting emissions". Davos Agenda. World Economic Forum. 2022, May 10. https://www.weforum.org/agenda/2022/05/electric-motorbikes-rwanda-ampersand/

506. Kabisa. "Go Electric". Go Kabisa. (n.d.). https://www.gokabisa.com/

507. Sky Energy Africa. "Electric Vehicles". (n.d.). https://www.skyenergyafrica.com/electric-vehicles/

508. Kaliwo, Khumbo. "As part of celebrating their 60th anniversary, TotalEnergies Marketing Malawi has launched three electric motorcycle charging points in Lilongwe.". @Times360Malawi. 2023, December 6. https://twitter.com/Times360Malawi/status/1732382823414145283/photo/1

509. Good Reads. "Incentives Quotes: Incentive structures work, so you have to be very careful of what you incent people to do, because various incentive structures create all sorts of consequences that you can't anticipate". Good Reads. (n.d.). https://www.goodreads.com/quotes/10531631-incentive-structures-work-so-you-have-to-be-very-careful

510. Merriam-Webster. "Society". Merriam-Webster. 2024, 11 January. https://www.merriam-webster.com/dictionary/society

511. Kenton, Will. "Economy: What It Is, Types of Economies, Economic Indicators". Economy. Investopedia. 2023, December 17. https://www.investopedia.com/terms/e/economy.asp

512. Bouchrika, Imed. "4 Types of Economic Systems in 2024: Which is Used by the World's Biggest Economies?". Education. Research.com. 2024, January 2. https://research.com/education/types-of-economic-systems

513. Pressbooks. "Types of Economic Systems". Pressbooks. (n.d.). https://pressbooks.howardcc.edu/soci101/chapter/13-2-types-of-economic-systems/

514. Rabie, Mohamed. "Economy and Society". Research Gate. 2016, June. https://www.researchgate.net/publication/304086270_Economy_and_Society

515. Rabie, Mohamed. "Economy and Society". Research Gate. 2016, June. https://www.researchgate.net/publication/304086270_Economy_and_Society

516. Merriam-Webster. "Culture". Dictionary. Merriam-Webster. 2024. https://www.merriam-webster.com/dictionary/culture

517. Ronaghi, Marzieh. Scorsone, Eric. "The Impact of COVID-19 Outbreak on CO2 Emissions in the Ten Countries with the Highest Carbon Dioxide Emissions". PubMed Central. 2023, June 13. https://www.ncbi.nlm.nih.gov/pmc/articles/PMC10281825.

518. Rabinowitz, Phil. "Section 8. Identifying and Analyzing Stakeholders and Their Interests". Community Tool Box. The University of Kansas. (n.d.). https://ctb.ku.edu/en/table-of-contents/participation/encouraging-involvement/identify-stakeholders/main

519. Bartik, Timothy J. Austin, John C. "Most Business Incentives Don't Work. Here's How to Fix Them". Brookings. 2019, November 4. https://www.brookings.edu/articles/most-business-incentives-dont-work-heres-how-to-fix-them/

520. Merriam-Webster. "Carrot-and-stick". Dictionary. Merriam Webster. (n.d.). https://www.merriam-webster.com/dictionary/carrot-and-stick

521. Geest, Gerrit De. Dari-Mattiacci. "The Rise of Carrots and the Decline of Sticks". Articles. University of Chicago. (n.d.). https://chicagounbound.uchicago.edu/cgi/viewcontent.cgi?referer=&httpsredir=1&article=5598&context=uclrev

522. Ollendyke, Dana. "Understanding Carbon Credits and Offsets". PennState Extension. Pennsylvania State. (n.d.). https://extension.psu.edu/understanding-carbon-credits-and-offsets

523. Legal Information Institute. "Cap-and-Trade". Wex Definition. Cornell Law School. https://www.law.cornell.edu/wex/cap-and-trade

524. Peterdy, Kyle. "Carbon Credits". Environment, Social, & Governance". CFI. (n.d.). https://corporatefinanceinstitute.com/resources/esg/carbon-credit/

525. Falk, Michael S. Klement, Joachim. "Green Carrots and Sticks: Incentivizing Climate Solutions". Drivers of Value, Economics, Future States, Investment Topics. Enterprising Investor. CFA Institute. 2021, June 23. https://blogs.cfainstitute.org/investor/2021/06/23/green-carrots-and-sticks-incentivizing-climate-solutions/

526. Shildrick, Tracy. "Incentives Quotes: Economics, when you strip away the guff and mathematical…". Poverty and Insecurity. Policy Press. 2012. https://books.google.mw/books?id=bgdbdjWNXskC&dq

[527]. Open Learn. "What is Politics?" Open Learn. (n.d.). https://www.open.edu/openlearn/society-politics-law/what-politics/content-section-2.1.4

[528]. Kagan, Julia. "What Is a Carbon Tax: Basics, Implementation, Offsets".Government & Policy. Investopedia. 2022, October 2. https://www.investopedia.com/terms/c/carbon-dioxide-tax.asp

[529]. The World Bank. "Carbon Pricing". Programs. The World Bank. (n.d.). https://www.worldbank.org/en/programs/pricing-carbon

[530]. The World Bank. "Carbon Pricing". Programs. The World Bank. (n.d.). https://www.worldbank.org/en/programs/pricing-carbon

[531]. Lai, Olivia. "What Countries Have a Carbon Tax?". Policy & Economics. Earth.org. 2021, September 10. https://Earth.org/what-countries-have-a-carbon-tax/

[532]. Gale, William G. Brown, Samuel. Saltiel, Fernando. "Carbon Taxes as Part of the Fiscal Solution". Research. Brookings. 2013, March 2013. https://www.brookings.edu/articles/carbon-taxes-as-part-of-the-fiscal-solution/

[533]. Parry, Ian. "Five Things to Know About Carbon Pricing". Finance & Development. International Monetary Fund. 2021, September. https://www.imf.org/en/Publications/fandd/issues/2021/09/five-things-to-know-about-carbon-pricing-parry

[534]. Parry, Ian. "Five Things to Know About Carbon Pricing". Finance & Development. International Monetary Fund. 2021, September. https://www.imf.org/en/Publications/fandd/issues/2021/09/five-things-to-know-about-carbon-pricing-parry

[535]. UCAR. "Effects of Climate Change on Ecology". Impact of Climate Change. Center for Science Education. (n.d.). https://scied.ucar.edu/learning-zone/climate-change-impacts/ecology

[536]. Department of Climate Change, Energy, the Environment and Water. "What are Wildlife Corridors?". Australian Government. (n.d.). https://www.dcceew.gov.au/environment/biodiversity/conservation/wildlife-corridors

[537]. Department of Climate Change, Energy, the Environment and Water. "What are Wildlife Corridors?". Australian Government. (n.d.). https://www.dcceew.gov.au/environment/biodiversity/conservation/wildlife-corridors

[538]. Dietz, Thomas. Shwom, Rachael L. Whitley, Cameron T. "Climate Change and Society". Annual Reviews. 2020, July. https://www.annualreviews.org/doi/10.1146/annurev-soc-121919-054614

539. Dietz, Thomas. Shwom, Rachel L. Whitley, Cameron T. "Climate Change and Society". Annual Review of Sociology. Annual Reviews.org. 2020, July. https://www.annualreviews.org/doi/10.1146/annurev-soc-121919-054614

540. Bundrick, Jacob. "Tax Incentives and Subsidies: Two Staples of Economic Development". Arkansas Center for Research in Economics. University of Central Arkansas. 2016, August 19. https://uca.edu/acre/2016/08/19/tax-incentives-and-subsidies-two-staples-of-economic-development/

541. Bundrick, Jacob. "Tax Incentives and Subsidies: Two Staples Of Economic Development". Arkansas Center for Research in Economics. University of Central Arkansas. 2016, August 19. https://uca.edu/acre/2016/08/19/tax-incentives-and-subsidies-two-staples-of-economic-development/

542. Libraries - University of Minnesota. "3.2 The Elements of Culture". University of Minnesota. 2016. https://open.lib.umn.edu/sociology/chapter/3-2-the-elements-of-culture/

543. Kamarck, Elaine. "The Challenging Politics of Climate Change". Articles. Brookings. 2019, September 23. https://www.brookings.edu/articles/the-challenging-politics-of-climate-change/

544. Kamarck, Elaine. "The Challenging Politics of Climate Change". Research. Brookings. 2019, September 23. https://www.brookings.edu/articles/the-challenging-politics-of-climate-change/

545. Falk, Michael S. Klement, Joachim. "Green Carrots and Sticks: Incentivizing Climate Solutions". Drivers of Value, Economics, Future States, Investment Topics. Blog. CFA Institute. 2021, June 23. https://blogs.cfainstitute.org/investor/2021/06/23/green-carrots-and-sticks-incentivizing-climate-solutions/

546. Good Reads. ""Be the change that you wish to see in the world."—Mahatma Gandhi". Change Quotes. Good Reads. (n.d.). https://www.goodreads.com/quotes/tag/change

547. Michigan Ross. "What Is Social Impact?". Business + Impact. Michigan University. (n.d.). https://businessimpact.umich.edu/about/what-is-social-impact/

548. Pringle, Patrick. Thomas, Adelle. "Climate Adaptation and Theory of Change: Making it work for you". Impact. (n.d.). https://ca1-clm.edcdn.com/assets/theory_of_change_briefing_note.pdf

549. Pringle, Patrick. Thomas, Adelle. "Climate Adaptation and Theory of Change: Making it work for you". Impact. (n.d.). https://ca1-clm.edcdn.com/assets/theory_of_change_briefing_note.pdf

550. Johnson-Woods. Wardlaw, Tait. "What is 'Theory of Change' and Why is it Important to Sustainability and Impact Initiatives?". Resonance Global. 2023, March 2. https://www.resonanceglobal.com/blog/what-is-theory-of-change-and-why-it-is-important-to-sustainability-and-impact-initiatives

551. GoodReads. "A nail is driven out by another nail; habit is overcome by habit.". Habits Quotes. Good Reads. (n.d.). https://www.forbes.com/quotes/2714/

552. Tett, Gillian. "Could behavioural nudges help us tackle the climate crisis?". Ft Magazine. Financial Times. 2021, September 15. https://www.ft.com/content/b6d36790-b171-417c-b593-b26900df3cd5

553. Elberg Nielsen, Anne Sofie. Sand, Henrik. Sørensen, Pernille. Knutsson, Mikael. Martinsson, Peter. Persson, Emil. Wollbrant, Conny. "Nudging and Pro-environmental Behaviour". Nordic Council of Ministers. 2016, October. https://norden.diva-portal.org/smash/get/diva2:1065958/FULLTEXT01.pdf

554. PETA. "10 Years Vegan Can Look Like This: Patrik Baboumian's New PETA Ad". Features. PETA. (n.d.). https://www.peta.org/features/vegan-patrik-baboumian-peta

555. Mohamed, Yomna. "Nudging for change: fighting climate crisis from our kitchens". Arab States. UNDP. 2022, August 8. https://www.undp.org/arab-states/blog/nudging-change-fighting-climate-crisis-our-kitchens

556. Mohamed, Yomna. "Nudging for change: fighting climate crisis from our kitchens". Arab States. UNDP. 2022, August 8. https://www.undp.org/arab-states/blog/nudging-change-fighting-climate-crisis-our-kitchens

557. Mohamed, Yomna. "Nudging for change: fighting climate crisis from our kitchens". Arab States. UNDP. 2022, August 8. https://www.undp.org/arab-states/blog/nudging-change-fighting-climate-crisis-our-kitchens

558. Mohamed, Yomna. "Nudging for change: fighting climate crisis from our kitchens". Arab States. UNDP. 2022, August 8. https://www.undp.org/arab-states/blog/nudging-change-fighting-climate-crisis-our-kitchens

559. Verel, Patrick. "'Nudges' Toward Climate Change Action May Discourage More Effective Activism". Science. Fordham News. Fordham. 2019, June 18. https://news.fordham.edu/science/nudges-toward-climate-change-action-may-discourage-more-effective-activism/

560. Verel, Patrick. "'Nudges' Toward Climate Change Action May Discourage More Effective Activism". Science. Fordham News. Fordham. 2019, June 18. https://news.fordham.edu/science/nudges-toward-climate-change-action-may-discourage-more-effective-activism/

561. Maia Films. "What if Everyone in the World Planted a Tree?". Ideas. BBC. 2020, February 26. https://www.bbc.co.uk/ideas/videos/what-if-everyone-in-the-world-planted-a-tree/p084ttpq

562. Igini, Martina. "10 Shocking Statistics About Deforestation". Crisis - BioSystem Viability". Earth.org. 2023 January 21st. https://Earth.org/statistics-deforestation/

[563]. Good Reads. "Quotable Quote: Watch your thoughts, they become your words; watch your words, they become your actions; watch your actions, they become your habits; watch your habits, they become your character; watch your character, it becomes your destiny.". Quotable Quotes. (n.d.). https://www.goodreads.com/quotes/8203490-watch-your-thoughts-they-become-your-words-watch-your-words

[564]. TEDx Talks. "The 5 Principles of Social Impact | Marian Spier | TEDxErasmusUniversityRotterdam". TEDx Talks. YouTube (video). 2017, December 18. https://www.youtube.com/watch?v=vsQJ2Y_F0ZY&t=3s

[565]. Good Reads. ""Dreams don't work unless you take action. The surest way to make your dreams come true is to live them.— Roy T. Bennett". Take Action Quotes. Good Reads. (n.d.). https://www.goodreads.com/quotes/tag/take-action

[566]. Natural Resources Manager. "Quotations" US Army Corps of Engineers. (n.d.). https://corpslakes.erdc.dren.mil/employees/perform/quotes.cfm

[567]. World Wide Fund for Nature. "FootPrint Calculator". World Wide Fund. (n.d.). https://footprint.wwf.org.uk/#/ https://www.wwf.de/themen-projekte/klimaschutz/wwf-klimarechner

[568]. Cozzi, Laura. Chen, Olivia. Kim, Hyeji. "The world's top 1% of emitters produce over 1000 times more CO2 than the bottom 1%". Commentaries. International Energy Agency. 2023, February 22. https://www.iea.org/commentaries/the-world-s-top-1-of-emitters-produce-over-1000-times-more-co2-than-the-bottom-1

[569]. US Environmental Protection Agency. "Greenhouse Gas Equivalencies Calculator". Energy and the Environment. Energy. US Environmental Protection Agency. (n.d.). https://www.epa.gov/energy/greenhouse-gas-equivalencies-calculator

[570]. Dark Matter Labs. "Designing our Future". Invitation Paper V.0.1. Desire. Dark Matter Labs. 2023. https://www.irresistiblecircularsociety.eu/assets/uploads/20230707-New-European-Bauhaus-Economy_Digital-version_DML.pdf

[571]. Zukav, Gary. The Seat of the Soul: 25th Anniversary Edition by Gary Zukav, Section: Study Guide, Chapter 6, Exercise: Every Emotion Is a Message, Quote Page 276, 2014,. Simon and Schuster. New York.

[572]. Adobe Express. "60+ quotes that will give you hope". Adobe. 2023, December 3. https://www.adobe.com/express/learn/blog/hope-quotes

[573]. Good Reads. "It is quite obvious that the human race has made a queer mess of life on this planet. But as a people, we probably harbour seeds of goodness that have lain for a long time waiting to sprout when conditions are right. Man's curiosity, his relentlessness, his inventiveness, his ingenuity have led him into deep trouble. We can only hope that these same traits will enable him to claw his way out.". Hope for the Future Quotes. (n.d.).

https://www.goodreads.com/quotes/7161890-it-is-quite-obvious-that-the-human-race-has-made

[574]. The Editors of Encyclopaedia Britannica. "Industrial Revolution". Britannica. 2024, January 29.
https://www.britannica.com/money/topic/Industrial-Revolution

[575]. The Editors of Encyclopaedia Britannica. "Industrial Revolution". Britannica. 2024, January 29.
https://www.britannica.com/money/topic/Industrial-Revolution

Made in the USA
Middletown, DE
14 May 2024

53988903R00144